LIVING CLOCKS IN THE ANIMAL WORLD

Publication Number 902
AMERICAN LECTURE SERIES

A Monograph in
AMERICAN LECTURES IN ENVIRONMENTAL STUDIES

Edited by
CHARLES G. WILBER
Professor of Zoology
Colorado State University
Fort Collins, Colorado

Living Clocks
in the
Animal World

By

MIRIAM F. BENNETT, Ph.D.

Chairman and Professor of Biology
Colby College
Waterville, Maine

CHARLES C THOMAS • PUBLISHER
Springfield · Illinois · U.S.A.

Published and Distributed Throughout the World by
CHARLES C THOMAS · PUBLISHER
Bannerstone House
301-327 East Lawrence Avenue, Springfield, Illinois, U.S.A.

©1974, *by* CHARLES C THOMAS · PUBLISHER
ISBN 0-398-02872-9
Library of Congress Catalog Card Number: 73-5573

*With THOMAS BOOKS careful attention is given to all details of
manufacturing and design. It is the Publisher's desire to present books that
are satisfactory as to their physical qualities and artistic possibilities and
appropriate for their particular use. THOMAS BOOKS will be true to those
laws of quality that assure a good name and good will.*

Printed in the United States of America
I-1

Library of Congress Cataloging in Publication Data
Bennett, Miriam F
    Living clocks in the animal world.
    (American lecture series, publication no. 902. A
monograph in American lectures in environmental studies)
    Bibliography: p.
    1. Biology—Periodicity.   I. Title.
QH527.B46   591.1   73-5573
ISBN 0-398-02872-9

For Peter and Andrew

FOREWORD

THIS FIRST VOLUME in the American Lecture Series on Environmental Studies is timely and appropriate. The study of biorhythms in animals (animal clocks) is a basic biological endeavor with important applied aspects.

Plants and animals show a distinct diurnal rhythm in their various activities. Cues from outside the organisms may force them into precise cycles. Invertebrate animals, lower vertebrates, mammals, subhuman primates and man exhibit a daily patterned cycle of rest and activity. This daily rhythm is called circadian (Latin "circa diem," about a day).

In many, there is a daily regular pulsation of body temperature, blood pressure, pulse rate, respiration, hemoglobin levels, and blood amino acid concentration.

Man lives on planet Earth which is subject to various rhythms — lunar, solar, seasonal. As a general rule, these natural rhythms are ignored by man in his social planning.

Recent studies in human biology show that "biological time of day appears to be hugely important in physiology. Time structure in the body may be almost as important as tissue."

Scheduling may well be a crucial factor in human well-being. What threat is posed by disrupting the deep-seated human circadian rhythm? We are not sure. It is certain that each species of animal does have a built-in rhythm of various functions. The rhythms depend on external (environmental) clues. It is long overdue that planning for human health and welfare acknowledge these inborn clocks as critical factors for *Homo sapiens.*

A recent study of nearly 2000 homicides in Dade County, Florida has shown that the murder rate "rose the day before a full moon, reached a peak at full moon and dropped back before a second peak at the new moon."[1] Virtually 90 percent of the homicides were committed during a full moon.

[1]Anon.: Murder and the moon. *The State Peace Officers Journal, 5:* 28-29, 1973.

Similar lunar influences were found for homicides in Cuyahoga County, Ohio. Internal "biological tides" in the human body have been suggested as possible violence triggering mechanisms.

Professor Bennett's book provides the basic biological data for a clearer understanding of the significance of biorhythms.

CHARLES G. WILBER

PREFACE

During the last twelve years, the reports of papers presented during several international conferences on biological timing have been published: The Cold Spring Harbor Symposium on *Biological Clocks* (1960), the Feldafing Summer School on *Circadian Clocks* (1965), the Friday Harbor Symposium on *Biochronometry* (1971), the Tihany Symposium on *Invertebrate Neurobiology: Mechanisms of Rhythm Regulation* (1973) and the International Society for the Study of Biological Rhythms meeting on *Chronobiology* (1973). Additionally, books focused on specific areas of biological timing are available. Among these are: Beck's *Insect Photoperiodism* (1968), Bünning's *The Physiological Clock* (1967), Cloudsley-Thompson's *Rhythmic Activity in Animal Physiology and Behaviour* (1961), Conroy's and Mill's *Human Circadian Rhythms* (1970), Harker's *The Physiology of Diurnal Rhythms* (1964), Palmer's *The Biological Clock. Two Views* (1970) and Sweeney's *Rhythmic Phenomena in Plants* (1969). Some of these publications as well as many studies of living clocks — both of the past and of the present — are pointed up in this book. I am convinced that familiarity with the literature and the investigators of any field of inquiry is demanded of its practitioners and is enjoyed by most of its spectators.

In this book, I have not attempted to review our knowledge of biological rhythms, in general, or that of timing mechanisms in large segments of the living world. The authors cited above — at least *in toto* — have done that and have done it well. Rather, I have tried to focus on the chronometry of particular animals — forms with which I have had some direct experience. I have attempted to underline the contributions of those animals to our fund of information about biological cycles. What have we learned from specific studies of specific animals? What major questions about their timing phenomena remain? How can we best attack those problems in our attempt to analyze, to synthesize and to understand the rhythms of those species and the rhythms of the living world as a whole?

ACKNOWLEDGMENTS

As a graduate student, I was introduced to animals' clocks by Professor Frank A. Brown, Jr. of Northwestern University. My thanks go to him for his guidance of my early work and for his interest in my studies of organismic timing which I continued at Sweet Briar College. I am also grateful to Professors Hansjochum Autrum and Maximillian Renner of the Institute of Zoology of the University of Munich, for the hospitality and aid afforded me during long-term stays in that institute.

I am especially happy to thank many of my students at Sweet Briar College — Dana C. Reinschmidt, Mary-Fleming Willis Finlay, Judith A. Harbottle, Carolyn B. Guilford, Jan Huguenin, Charlene Reed and Joan H. Spisso — for their technical assistance, discussions, arguments, criticisms and advice which have contributed immeasurably to my studies of animals' clocks, to my understanding of the temporal behavior of animals and to my delight in learning and teaching about living clocks.

It is also a great pleasure to acknowledge the contributions of Dean Catherine S. Sims of Sweet Briar College who encouraged me to write this book and who aided the project in many, many ways. Financial help for my own studies and for the preparation of the book has come from the Office of Naval Research, the National Science Foundation, the Sigma Xi Club of Lynchburg, Virginia, the Kampmann Award of the Sweet Briar Alumnae Association, the Committee on Faculty Research, Sweet Briar College and the Anonymous Donor's Science Fund of Sweet Briar College.

The persons who read and corrected my manuscript — in particular Naomi B. Erdmann and Dorothy Vickery — deserve many thanks. But, they share no blame for errors or points of confusion. Those are all mine. I also wish to thank Julia S. Child and artists of the Frank Wright Studio of Lynchburg, Virginia for their drawings and Cecille Harvey for her typing.

Authors whose figures have been used are cited in the legends. My thanks go to them and to the following for permission to publish illustrations which have been copyrighted:

Academic Press, Inc., New York

American Association for the Advancement of Science, Washington

The American Midland Naturalist, Notre Dame

American Physiological Society, Bethesda

The Biological Bulletin, Woods Hole

Cold Spring Harbor Laboratory, Cold Spring Harbor

Cornell University Press, Ithaca

Macmillan Journals, Ltd., London

The National Academy of Sciences, Washington

North-Holland Publishing Company, Amsterdam

Springer-Verlag, Berlin-Heidelberg-New York

The Society for Experimental Biology and Medicine, Utica

The University of Chicago Press, Chicago

The Wistar Institute of Anatomy and Biology, Philadelphia

July, 1973   Miriam F. Bennett

CONTENTS

LIVING CLOCKS IN THE ANIMAL WORLD

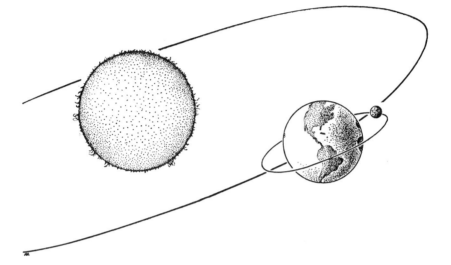

Chapter 1

LIVING CLOCKS
PAST AND PRESENT

LOREN EISELEY, himself very much a student of life in terms of time, believes ours to be "... the most time-conscious generation that has ever lived." (Eiseley, 1969, p. 5). We have come by this reputation simply and honestly, for we merely reflect the billions of years of evolution of biological timing that have come before us. Time-consciousness is undoubtedly an attribute only of man, but having the capacity to live in time with changes of the sun and the moon, the tides and the seasons is characteristic of all life on this earth, and probably has been so since that life came into being.

Man's realization that he is not the only timekeeping species is ancient. Androsthenes, an officer who accompanied Alexander the Great on his explorations, observed the daytime rising and the nighttime falling of the leaves of the tropical Tamarind tree. Similar notes about changing positions of leaves are found in the first century writings of Pliny, the Elder, and in the thirteenth century writings of Albertus Magnus. These changes are now referred to as "sleep movements," for in the seventeenth century, Linnaeus described the nocturnal positions of leaves as plant sleep. In the eighteenth century, De Mairan, an astronomer most interested in the rotation of the earth, and the botanists, Duhamel and Zinn, were able to report that changes in light intensity of the plants' surroundings did not cause the sleep movements of the leaves, since their rising and falling continued for some period of time even when the plants were kept in constant darkness or under constant artificial light. European botanists continued to lead as investigators of biological timing during the nineteenth century when they were joined by Charles Darwin, who described temporal changes in clover and wood sorrel as well as in earthworms, and by the Viennese zoologist, Kiesel, who recorded day-night changes in the retinal pigments of arthropods.

So, by the beginning of the twentieth century, examples of both plant and animal clocks had been recorded in the scientific literature. What are these biological or living clocks? Biological clocks are inherited, cellular mechanisms with which organisms mark off or measure the passage of time cyclically. The ticking or beating of such precisely integrated collections of molecules — whatever they may be — allows living systems to vary the intensity or the level of many diverse functions in regularly repetitive fashions. The rhythmically recurring changes, *e.g.,* the positions of the leaves of plants, the positions of the pigments of the eyes of shrimp, the rates of crawling of earthworms, which are regulated by the clock, *but are not identical with it,* continue for some time even when the organisms are removed from their natural environments, and are maintained under laboratory conditions of constant light intensity, constant temperature and constant humidity. The frequencies of those changes or the periods of time between the occurrence of similar events or levels of the cycles under the constant conditions are very close to those measured while the organisms are exposed to changes in natural environmental factors such as light and temperature which occur fairly regularly in their real worlds. Therefore, one can state that by virtue of cellular clocks, rhythms of some organismic activities persist under the unvarying and thereby, the atypical "constant conditions" of the laboratory.

The present-day student of the clocks of animals should stop briefly during a rapid review of the history of studies of biological timing to consider the investigations of Pfeffer, much of whose work was on the clocks of plants, especially of beans and marigolds. At first, he doubted the validity of claims of the continuance or persistence of sleep movements under constant conditions. He thought that perhaps light which had leaked into caves and darkrooms used for the eighteenth century experiments had induced the daytime positions of the leaves. So, he repeated some of those experiments. After his own observations supported the findings of Duhamel and Zinn, he, also, was convinced of the reality of biological changes which persisted at frequencies similar to those of changes in the plants' normal outside worlds even when the organisms were in the laboratory and were deprived of physical variations, *e.g.*, light to dark, which might function in the natural environments of the plants as temporal cues or time signals.

An endogenous component of organisms' 24-hour cycles was therefore recognized and emphasized by the early twentieth century, for Pfeffer concluded that the frequencies of plant rhythms depended upon an internal, organic mechanism. This mechanism, independent of environmental factors, was believed to generate the periods or durations of persistent periodic phenomena, other aspects of which, amplitude and phasing, were influenced and regulated by changes in the physical surroundings of the plants. Here, basically, is the view of the nature of persistent biological rhythms of plants, animals and microorganisms held by most students of biochronometry today.

The second view of the nature of the basis of organic timing was also espoused in the early twentieth century, again by a botanist, Stoppel. She and her present-day supporters believe that living clocks demand information from their physical environments to be able to generate the frequencies of their cyclic activities. Even under "constant laboratory conditions" subtle, cyclic geophysical factors are not screened out and therefore are available to organisms as time cues. According to this hypothesis, physical factors not only affect the amplitude and form of persistent rhythms, but they are also necessary for the maintenance of the lengths or periods of the cycles. Pfeffer, Stoppel, their forerunners and their contemporaries did indeed lay the foundation for the newer attacks on biological clocks which attracted the attentions of more and more biologists of many different specialties during the 1930's. That attraction has increased, and in terms of the number of investigators active in biochronometry, ours certainly is the most time-conscious generation of biologists that has ever lived.

Erwin Bünning of the University of Tübingen whose own writings (1960 and 1967) include excellent historical surveys of the study of biological clocks, is one of this century's leaders in the field. His earlier work focused on rhythms of both plants and animals, although that of more recent years has concerned only plants. His analyses of the effects of temperature and light changes on cycles are outstanding. His emphasis on the relationship between persistent rhythms and photoperiodic effects have stimulated significant recent studies (*e.g.*, Pittendrigh, 1963). Bünning's hypotheses of the oscillatory nature of the internal timing mechanism have prompted the development of models, mathematical theories and engineering parallels which attempt to explain the basic workings of living clocks. Observations and experimental results from biological laboratories all over the world confirm and extend many of his original ideas.

Before the Second World War, Hans Kalmus of the University of London, Karl von Frisch of the University of Munich and the late Orlando Park of Northwestern University were leading the zoologists in their observations and analyses of animal clocks and of the ecological and behavioral significance such timing mechanisms have for their possessors. Curt Richter of the Johns Hopkins University, Nathaniel Kleitman of the University of Chicago and Franz Halberg, now at the University of Minnesota, were among the first to investigate precisely, periodic changes in human beings and to emphasize their importance in medicine and its practice. The reports of an intensive study of living clocks by Frank Brown and his students at Northwestern University and the Marine Biological Laboratory at Woods Hole, Massachusetts, began in 1948. That program continues, and as is to be seen in later chapters of this book, its publications contain records of some of the most amazing, provocative and enigmatic aspects of organismic timing. Brown now leads the school of investigators who believe, as did Stoppel, that cyclic, subtle geophysical factors are used by organisms to time their persistent rhythms.

During the 1950's the late Gustav Kramer of the Max Planck Institute in Wilhelmshaven, and von Frisch proved that animals so different as starlings and honey bees are able to move to the right place at the right time because of an intricate association of spatial and temporal behavioral patterns. Their works gave rise to the concept of clock-compass reactions or time-compensated orientation. In the same years, Jürgen Aschoff of the Max Planck Institute in Erling-Andechs and Colin Pittendrigh, now at Stanford University, came into prominence. Between them and their two groups of students and associates, they have probably studied physiological clocks in every major type of living system — microorganism, plant and animal. Their experiments have been beautifully conceived and executed. Their analyses of many parameters of rhythmicity and their variations are fundamental to biochronometry. Aschoff and Pittendrigh with Bünning lead the present-day adherents of the endogenous theory of biological timing, the idea supported by Pfeffer in the nineteenth century.

Yet now — in the second half of the twentieth century, most of the basic and fascinating questions and problems regarding living clocks remain. Their solutions will demand the continued enthusiasm and energy of a great many time-conscious investigators of our generation and of the generations to come. We must learn the actual structures and workings of biological clocks. What are their components? How do they function? How are they identified? What relationships do the clocks and their parts have to the physical environment of our earth? Are they adaptive to life on this planet? What relevance may they and knowledge of them have for us, a time-bound species? Partial answers to many of these questions have come to us from our work on animal clocks. However, a general discussion of the questions and answers which we have from studies of many forms of life is appropriate and necessary at this point to illustrate and to emphasize the principles, the theories and the vocabulary with which the student of biochronometry works.

That living clocks are cellular in nature has already been stated. Granted, only a part of the molecular population of cells or groups of cells may actually be engaged in generating the temporal outputs of living timepieces, but those molecules are parts of cells. To date, subcellular or acellular organic systems have not been shown to function as biological clocks. It has also been pointed out that the rhythmic processes, the crawling of earthworms and the sleep movements of leaves, which are regulated by the living clocks, persist in time with physical changes in the worms' or trees' natural habitats even after the organisms have been placed under "constant conditions" in the laboratory.

The periods, or their reciprocals, the frequencies, of the persistent rhythms are basically the same as those of the geophysical cycles of our surroundings. And these cycles, of course, depend upon the relative positions and movements of our earth, our moon and our sun. The majority of the persistent organismic cycles known and well studied are 24 hours in duration — the time of one rotation of the earth on its axis. These are usually called solar-day or circadian rhythms, but some authors continue to refer to them as diurnal rhythms. Cycles which recur at 12.4-hour intervals are tidal or primary lunar ones, while those of 24.8 hours are lunar-day rhythms. The frequencies of both these types of lunar cycles are the same as those of cycles of the environment which reflect movements of the earth relative to the moon, and those of the great air and water masses of the earth which are caused by the position of earth *vis a vis* moon. Lunar-monthly or 29.75-day rhythms have also been well documented for many organisms. A monthly rhythm may be the consequence of the organisms' simultaneous functioning at solar- and lunar-day frequencies. Lunar-monthly rhythms allow organisms to live in time with the orbiting of the moon around the earth. Least familiar and least studied are persistent annual cycles, those of approximately a year's duration, the time of the revolution of the earth around the sun. However, a few examples of yearly cycles of living systems have been recorded, and some are to be discussed in this book.

In the nineteenth century, De Candolle, a French botanist who investigated the sleep movements of *Mimosa*, clover and bean plants, noted the periods of some of those cycles to be a bit shorter than 24 hours. We now emphasize that the periods of most organismic rhythms expressed under constant laboratory conditions are very close to, but not identical with, the periods of geophysical cycles. Such periods or frequencies are the free-running ones, *i.e.*, they are seen in organisms living under the so-called constant laboratory conditions where no obvious time signals are conveyed by observational or experimental procedures. Free-running periods may be, and often are, longer or shorter than are the periods of the organismic cycles when the plants or animals are cued or are entrained by regularly repeated changes in their surroundings — either natural or experimental.

Thus, under free-running conditions, the persistent rhythms often run slightly out of phase with the actual day-night, lunar and annual rhythms of the earth. However, the periods of the cycles induced by living clocks vary within only narrow limits, *e.g.*, roughly 19 to 30 hours for circadian or solar cycles, and in a beautifully precise and highly adaptive manner can be attuned to the rhythms of the natural world by common and reliable changes of our normally varying physical environment.

Thus far I have focused upon the cellular timing mechanism, the living clock itself, but in doing so I have, necessarily, referred also to its hands, its indicators or the periodic organismic changes which persist under constant conditions. Again, one finds a European botanist of the nineteenth century, Sachs, to be the first investigator to emphasize this principle of biochronometry: living clocks and their hands are not identical with one another. He pointed out that periodic leaf movements, or the hands of the clocks of some plants, are dependent upon the periodicity of an "entire complex of processes." Sach's "entire complex of processes" may be equated with the cellular clock. It is imperative that we always distinguish between the clock and the hands, for their functions and characteristics vary, and several different sets of hands may be attached to the same set of clockworks. The common fiddler crab of our Atlantic shores shows persistent rhythms of both color change (Chapter 2) and locomotor activity (Chapter 3). One set of hands of its clock is color change; the second indicator of the clock is locomotor activity. Additionally, the anatomical and physiological attachments between the cellular timing mechanism and its indicators must be considered. Those tie-ups are often called mediating or regulatory pathways or transmission systems. In the fiddler crab, the mediating pathways between its biological clock and color change are primarily hormonal, while regulation of its cycles of locomotion depends upon the crab's nervous system. A variety of specific indicators and their regulatory systems are to be described and discussed in later chapters of this book.

And, almost chapter by chapter, these questions will arise: what is the true nature of living clocks and how do they function? As has been and will be stated repeatedly, we can not describe either the detailed anatomy or physiology of the clocks. We do not know exactly what they are or how they act. In an excellent recent review, *The Biological Clock. Two Views.* (Palmer, 1970), Frank Brown and J. W. Hastings each states and defends one of the two views. Brown deals with the hypothesis of environmental timing of the clock. Hastings concerns himself with that of a purely endogenous and independent clock. Both views embrace a physicochemical cellular mechanism which, according to Hastings, is fully autonomous, and which, according to Brown, must have informational input from the physical environment to be able to maintain its average precision and basic functioning. That input consists of fluctuations of geophysical factors of the earth and its sphere of influence which permeate the "constant conditions" of the laboratory, and repeat at solar and lunar frequencies. The organism "locks" its clock to those regular changes. What the factors are is also unknown, but geomagnetism and particular components of background radiation are leading candidates.

Naturally, any valid theory of the nature of living clocks must explain all characteristics of the basic timing mechanisms including two of their most amazing physiological ones: virtual temperature-independence or temperature-compensation through a wide range and insensitivity to chemical disruptions as great as some effected by cyanide, antibiotics and metabolic hormones. Ever since attacks on animal clocks began, investigators have marveled at the capacity of the cellular clocks to maintain their usual frequencies between approximately 5° and 30° C. They are neither slowed down by cold nor speeded up by heat, deviating, in those respects, from many biochemical systems which do have temperature coefficients of 2.0 to 4.0. For a 10° C rise in temperature, the rates of most chemical reactions are doubled, tripled or quadrupled.

The temperature-independence of living clocks remains one of our major enigmas. Will it be explained on the basis of a biochemical temperature-compensatory system, of an inherent property of biochemical oscillators, or as a natural consequence of the clock's being regulated by fluctuating geophysical factors whose effects are the same at 5° C as at 30° C? And how can the cellular clocks beat out their normal frequencies when vital metabolic steps, *e.g.*, enzyme or energy production, are inhibited or speeded up by chemical agents? Is absolutely balanced "activation followed by inhibition" (Hastings, 1970. p. 79) another unique property of physicochemical timing phenomena? Or can the clock which has been treated with drugs still sustain its normal reactions to changes in the levels of subtle geophysical factors, and thereby continue to receive and to generate information regarding the time of day, month and year?

The analyses of the effects of temperature and drugs on living clocks, several of which are discussed in later chapters, may very well help us decide whether the cellular timepieces under study are fully autonomous or are dependent upon exogenous signals. Although the nature of biological clocks is most often thought of in terms of either endogenous or exogenous, the possibility that both autonomous and environmentally modulated cellular clocks exist should be stressed. Biologists would certainly not be shocked to find timing redundancies in organisms and timing differences among organisms. Variations superimposed upon fundamental likenesses — anatomical, physiological and behavioral — are the rule in the living world of this earth.

Variations among living clocks are even more conceivable when one looks at the tremendous spectrum of types of organisms in which timing reactions have been found. Algae, protozoans, the plants and animals of most phyla and habitats are represented in our list of rhythmic species. Most of us are perfectly willing to suggest the ubiquity of cellular clocks. But are they all the same? The hands of the clocks and their mediating pathways vary tremendously. Cycles of rates of cell division, gamete production, hatching, oxygen-consumption, biosynthesis and bioelectric changes exist in many organisms. Other forms show rhythms of locomotion, movements of their parts or changes in pigmentation. A great number of these periodic changes are clearly overt and can be measured or seen obviously day after day. The circadian cycle of hatching of *Drosophila*, the fruit fly, and the tidal cycle of running of *Uca*, the fiddler crab, are overt ones.

Fewer organismic cycles are statistical ones, discernible only after the averaging of data of many hours, days or months. Mean solar-day cycles of the oxygen-consumption of potato tubers, based on data recorded for every hour of a year, have always shown three peaks — one at 7:00 AM, one at noon and the third at 6:00 PM (Brown, 1969). The tendency of mud snails to turn varies during solar and lunar periods (Chapter 6). Based on hourly observations made through a three-month period, the average circadian cycle of snails' turning had a maximum at noon with minima at 5:00 AM and 7:00 PM. Some organismic cycles may be direct reflections of the ticking of the cellular clocks. Examples are changes in the levels of respiration or the production of enzymes. Other rhythms must be mediated by chemical or nervous activities standing between the clock and its hands. Color change and locomotory rhythms are examples of these. We students of biochronometry have started to operate as comparative biologists. That approach should be continued and intensified.

The comparing and contrasting of rhythmic phenomena should not be limited to an interspecific level. Differences among cycles of individuals of one species; changes in rhythms during ontogeny; variations in them from day to day and month to month in mature organisms all should be looked into systematically. Aschoff (1960 and 1965) emphasized individual variations in frequencies under free-running conditions, and corresponding variations among ratios of animals' activity and rest. Commonly, one also finds that the rhythms of individuals of the same species exhibit different amplitudes. The amplitude of a cycle is the difference between its maximal and minimal levels. In one organism, amplitudes and average levels of a rhythm may vary "spontaneously" or after experimental treatment. The clock of the fiddler crab is not slowed down by some low temperatures, but the amplitude of its circadian cycle of color change is affected by the cold. Phasing, the temporal relationship between a specific organismic event, *e.g.*, the onset of running of an animal, and a specific time, *e.g.*, 6:00 PM, may also vary, and can be manipulated in diverse ways.

This flexibility of phasing, another major characteristic of persistent rhythms, assures the fitting of organisms' cycles to rhythms of their immediate worlds in manners adaptive to the organisms and exciting to students of biochronometry. Flying squirrels which lived under natural daylight conditions began their running about the time of sunset which varies through the year. Under laboratory conditions, that onset of activity could be set to many points in the solar period by decreasing the light intensity to which the squirrels were exposed (DeCoursey, 1960). Systematic investigations of phasing and phase-shifting are elucidating properties of those phenomena, and are teaching us much about the synchronization of events of organismic and environmental cycles by time cues, Zeitgebers (literally — time-givers) or entraining agents.

If, while in its naturally varying physical environment, an

organism's cycle were to run at a frequency different from that of the solar cycle, as many circadian cycles do under constant conditions, and could not be entrained to events of the 24-hour day, its possession might become a greater liability than asset. This situation is not probable, however, for as flying squirrels and many other organisms have shown us, the phases of persistent rhythms can be set at different points in terms of actual time by particular changes or perturbations.

A cockroach can be caused to run at its peak level at most any hour by exposing it to a light to dark change at an appropriate time (Harker, 1964). The phases of the insect's rhythm will then maintain their approximate relationships with real time even under constant conditions until new Zeitgebers are presented. In its natural habitat, the roach will probably run maximally shortly after sunset even though the solar time of sunset varies through the year. If for any reason, the insect does not experience light to dark changes for a period of days, it can still be expected to be most active during early nighttime. Its cellular clock continues to direct its periodic changes in levels of its activity — approximately in time with those of its surroundings. Many living clocks, whether they be purely autonomous, environmentally cued or both, are not just luxuries. They are necessities for success in a varying world.

Let us realize that adherents of both views of the nature of the internal clock find phase-shifting of persistent rhythms explicable. The endogenous hypothesis does not assert that living systems are insensitive to changes in environmental factors — obvious or subtle ones. Proponents of this theory do claim, however, that the clock runs independent of time cues from such factors, as it does in a free-run. But when the organism is in its natural environment, its cellular clock acts with changes in its surroundings — resulting in the organismic rhythm's being beautifully in time with the physical world.

On the other hand, those who support the exogenous hypothesis do not believe that the hands of the clock are locked to particular changes in the geophysical world. The cellular clock, itself, is so locked, and is therefore modulated by these regular changes. In addition, the hands or indicators of the clock can also be affected by environmental variations. Phases of its cycles can be shifted relative to states of the internal clock, and thus to real time, by a variety of perturbations in organisms' surroundings. Several discussions of phase-shifting are included in following chapters where its great value to specific animals will be stressed.

Not all changes nor changes at all times are equally effective in entraining rhythms or shifting phases. Organisms possess rhythmic fluctuations in sensitivity to Zeitgebers. Light shocks of 10 minutes given to flying squirrels which are otherwise maintained in constant darkness shift the phases of their activity cycles only if presented during the animals' subjective night. Ten minutes of light during the squirrels' subjective daytime have no effects on phasing (DeCoursey, 1960). Are those variations in the reactions to perturbations properties of the cellular clock? Its hands? Its mediating pathways, or combinations of all those components?

Once more, the situation may vary from species to species. In the discussions to follow, the Zeitgebers most often considered will be changes in light intensity and ambient temperature. These are among the most obvious perturbations in organisms' natural habitats, and are the ones which have been most thoroughly studied in the laboratory. Sound, mechanical pressures and social cues are also effective phasing agents in some animals. Our information about them is unfortunately meager. All these findings which indicate flexibility superimposed on reliability of the timing of vital events add weight to the notion that living clocks and the phenomena cued by them — persistent rhythms, time-compensated orientation and photoperiodism — are adaptive to living systems on our earth, and undoubtedly have been of selective value during the billions of years of organic evolution which have taken place on it.

It is strange that Charles Darwin who observed and described persistent rhythms in both plants and animals well after the publication of his *opus magnum, The Origin of The Species,* did not stress the possible selective value of cyclic organismic activities in his monographs. However, one of his points regarding the sleep movements of plants which may have been his clue of physiological adaptation has recently been strengthened by Bünning's (1971) investigations. Darwin suggested that the surfaces of the leaves in their night positions — folded or drooping with their long axes at right angles to the ground — were not in so much danger of absorbing moonlight, which might be detrimental, as they would be if the usual daytime positions were maintained.

Very recent investigations have proved that exposure to light, even of low intensity, during subjective night is a powerful phase-shifting procedure which affects photoperiodic reactions in adverse ways. Sleep positions protect the plants against such shifts. Perhaps, some rhythms which do not appear to confer biological advantages; and one thinks immediately of lunar cycles of activity and metabolism in fresh water and terrestrial animals and plants as examples, will be found to be adaptive when we know more about the details of the lives of those organisms. Finally, as with all biological properties, some — in this case, periodic phenomena — may not be adaptive. They may be neutral and merely relics whose development and maintenance are contained within the genomes of particular species.

Nevertheless, the majority of organismic rhythms do help living systems fit themselves to their surroundings. That fitting or existing in temporal synchrony with their environments, which allows certain animals to be in the right place at the right time or to do the right thing at the right time, is another focal point of this book. Because of its living clock and the clock's precision, an individual does not have to wait to be pushed passively to its proper place or chore by a physical change. It anticipates, and is ready for the trigger or Zeitgeber, and even if that does not come for several cycles, the organism can function in its proper rhythm. And, as the real time of cosmic events — sunrise, low tide, sunset — which entrain animals' cycles, varies through the day, the month and the year, the phasing of the organism's rhythms is adjusted accordingly. As stated earlier, that adjustment is made by the reliable cellular clock and its mediating pathways which together can regulate the various hands of the clock in ways which are certainly relevant to the lives of the forms which possess living chronometers.

Their relevance to us, the members of the only time-conscious species, is made more obvious almost daily. Our newspapers describe how intercontinental flight crews and passengers feel and act when their internal times are out of joint with external times. The cartoonists contribute to the publicity of biological clocks. Charlie Brown's Snoopy is well aware of his "stomach clock" and the agony he feels when suppertime is late. Readers of *The New Yorker* smile knowingly at the husband deplaning in Europe who admonishes his wife to wait to shop until her inner clock has become adjusted. The concern of NASA for spacemen who must function for days at a time without the configuration of geophysical time signals to which life on earth is accustomed should be and is registered often. Medical, psychiatric and safety-engineering applications of the knowledge of human cycles are often discussed. In the daily press and in the "better magazines," one can read of how human beings react and how their behavioral and physiological cycles run when they spend weeks alone in caves or in experimental bunkers under constant conditions of light, temperature and humidity.

At the present time, Aschoff and his associates are conducing fascinating investigations of the efficacy of social interactions and social cues as Zeitgebers for persons living in underground bunkers where physical conditions and schedules can be rigorously controlled. Can going to work or going to lunch entrain one's circadian cycles to the natural 24-hour day? Biologists, social scientists and laymen will want to follow the progress of Aschoff's studies. Most of us will probably also follow the discussions of the frightening and frustrating rapidity of the changes to which "modern man" is expected to adapt. Should the pace of human life be geared technically to a tempo that could entrain the actual cycles of man's physiology? Could such conscious entrainment protect us from desynchronization and its possible stressful effects?

How living clocks are related to the endeavors of zoologists of many specialties is also to be pointed out in the discussions which follow. Suffice it to say now that no one of us can interpret fully the results of any of our investigations without keeping rhythms and rhythmic activities in mind. Can a zoological experiment ever be repeated exactly if the animal under study is not the same system from minute to minute, hour to hour or day to day? Investigators should be well apprised of time as an observational and experimental variable, and must keep time in mind as they design and execute their studies.

Further, I have thought it important to emphasize in this book the gaps which exist in the stories of the clocks of animals which I have tried to tell. Research problems abound. The beginner and the established investigator both have the opportunity to contribute something new to biochronometry. Answering questions about animal clocks can excite the intellect, can teach more of the science of life and can provide information for medicine, agriculture and bioengineering. Perhaps greater knowledge and greater understanding of the many clocks of animals can, in addition, make all of us time-conscious human beings a little less lonely and a little less frustrated in our rapidly cycling universe. It may be reassuring to know that all life is basically rhythmic.

Chapter 2

THE FIDDLER CRAB AND ITS COLOR CHANGE

In April, 1954, an article entitled "Biological Clocks and The Fiddler Crab" appeared in *Scientific American*. In that paper, the author, Frank Brown, introduced his readers to the study of biochronometry by reminding them of the ability of many organisms, including man, to keep track of the passage of time without using conventional clocks, tide tables or calendars. He presented examples of well-known indicators of biological timing: the swarming of the Atlantic fireworm and the Pacific grunion; the running of insects and mice; and the time-compensated navigation of honey bees and starlings. He also explained that his own interest in organismic timekeeping was aroused by learning that cold-blooded animals apparently maintain 24-hour cycles of physiological processes regardless of the temperature at which they live. The biological clock of those animals seemed to him to be independent of temperature, at least in the organisms' physiological ranges. Their clocks did not gain time when the animals were warm, and did not lose time when they were cool.

Those observations do not parallel the results of all studies of the effects of temperature on the rates of chemical reactions which underlie physiological processes. Many of those reactions generally are speeded up by higher temperatures and are slowed down by lower ones. For his early investigations of the temperature characteristics of animal clocks, Brown chose *Uca*, fiddler crabs, cold-blooded forms, whose coloration was known to vary with times of the solar cycle even when the crabs lived in the laboratory under constant light intensity and temperature. The remainder of the 1954 article reviewed other fascinating aspects of biochronometry which had been learned by that time from the studies of fiddler crabs, small inhabitants of many seashores of the world. Now, almost two decades later, much more must be added to the list of contributions of those animals to the field of biological rhythmicity.

The crabs are called fiddlers because of the movements of the one especially large claw or cheliped of the males. Their large appendages are moved back and forth in a beckoning motion which reminds one of the movements of the bow arm of a violinist or fiddler. Most *Uca* live on seashores where they burrow into the sand or mud between the tide lines during high tide. As the water recedes from the area of an animal's burrow, the crab emerges and follows the water seaward, stuffing its mouth with mud and organic debris picked up in its claws from the beach. Members of some species of *Uca* — *pugnax* and *pugilator* — run and feed at low tides in large groups which can be driven about on the tidal flats by collectors or other interested observers. Soon after low tide, the crabs start to move up the beach where they dig new burrows and plug their openings with sand. They remain in the burrows during high tide, and until the water moves down the flats with the approach of the next low tide. There is no question that these semiterrestrial crabs live in time with the tides.

And, for fiddler crabs of the temperate zones, there is no question that the animals also live in time with the seasons of the year. The warmer periods are spent feeding with the tides, maturing and reproducing. In the winter, the crabs spend most of their time in their burrows where they exist at very low metabolic rates. But, are phases of the solar cycle also reflected in the behavior of *Uca*? Females that are carrying eggs come out of their burrows to feed only in the evening; the larvae hatch from their eggs only at dusk. Although people who live near *Uca*-inhabited flats on Cape Cod have reported that during some nights they are kept from sleep by the noise of groups of fiddler crabs moving through the beach grasses, and a few field observations support this hint of some nocturnal activity — collecting trips scheduled for low tides of the evening or night are not very successful. Few fiddler crabs are seen out on the beaches at those times. Extensive comparative studies of the behavior of many species of *Uca*, of both temperate and tropical zones, have also shown that with one or two exceptions, those crustaceans are diurnally active· (Crane, 1944 and 1958).

The first persistent cycle described for fiddler crabs was a very obvious 24-hour rhythm of color change. In 1912, Megušar published the results of his observations of that cycle in *Uca pugnax*. The animals with which Megušar worked in his laboratory in Vienna had been taken to him from Woods Hole, Massachusetts, by a colleague who had been traveling in the United States. Even after the long ocean voyage, the crabs' periodic change in coloration was apparent. Early in the morning, the legs of the crabs were light—almost white; by midday, the appendages had become dark—almost black; but by late evening and through the night, they were light again (Fig. 1). Those changes were observed day after day in animals that lived under natural day-night changes in illumination. While some of the specimens were exposed to reversed lighting of an intensity of 50 ft c from 8:00 PM to 7:00 AM and darkness from then until 8:00 PM, Megušar saw that the phases of their color change cycle were also reversed, as illustrated in Figure 2. Then, the crabs had very dark legs in the mornings and evenings, and had very light ones during midday. The Viennese investigator thus had also proved that a specific event or phase of an animal's persistent cycle is not locked

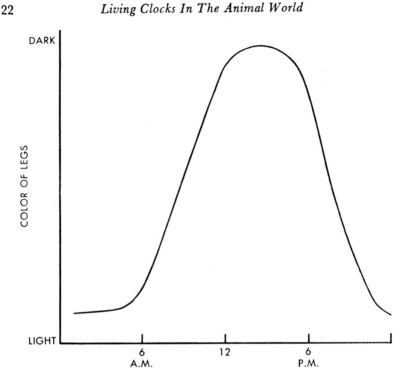

DARK

COLOR OF LEGS

LIGHT

6
A.M.

12

6
P.M.

TIME OF DAY

Figure 1.   Variations in the color of the legs of fiddler crabs during one solar-day. The animals were exposed to natural day-night changes in the laboratory.

to a particular real time of the circadian period. Therefore, flexibility of phasing of organismic rhythms was known for the fiddler crab from the very beginning of the study of its daily cycle of color change.

A comparable rhythm for *Uca pugilator*, also from Woods Hole, was described in 1937 by Abramowitz, and his report offered information concerning the possible regulation of the color change cycle in fiddler crabs. The legs of calico-backs, as *Uca pugilator* are often called, were also found to be dark during the day and light during the night when the animals were exposed to normal day-night changes and when they lived in complete darkness. The explanation offered was this: the con-

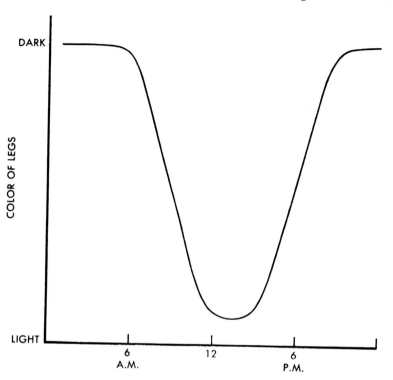

DARK

COLOR OF LEGS

LIGHT

6
A.M.

12

6
P.M.

TIME OF DAY

Figure 2.   Variations in the color of the legs of fiddler crabs during one solar-day. The animals were exposed to darkness by day and to light by night.

centration of a hormone, which is produced by organs in the eyestalks of the animals, and which causes the dispersion of black pigment in the color cells of the legs of the crabs, is higher during the daytime than during the nighttime. When the black pigment is dispersed, the legs take on a general darkened appearance. Conversely, when the hormonal concentration is lower and the dark pigment is concentrated in small points within the color cells of the legs, the appendages are much lighter in hue. Abramowitz found that if only one eyestalk were removed from a crab, its rhythm of color change continued. He claimed that after both eyestalks were amputated, the cycle was abolished.

Later studies by other investigators caused us to modify that latter statement substantially; but they, too, support Abramowitz's conclusion that hormones are primary mediators of the color change rhythms of fiddler crabs.

How can we explain the temporal variations in the amounts of the chromatophorotropins or color change hormones in the body fluids? It has been believed that the hormonal content of the glands which actually produce those chemical messengers is roughly the same throughout the circadian period, and that the hormones are released from those glands only periodically. What, then, triggers the periodic release of the color change hormones? How do the clockworks of the fiddler crab cause the changes in titers of those hormones? Complete answers to those questions are not yet available. Nevertheless, suggestions for partial answers and evidence of temporal variations in the potency of at least some chromatophorotropins have been recorded in the literature, and should be helpful in explaining hormonal mediation of rhythmic color change in *Uca* and other organisms.

Early in his investigations of fiddler crabs' cycles, Frank Brown postulated that to be adaptive, a living clock must be virtually temperature-independent or have a temperature quotient very close to zero. If such a temperature relationship did not exist, the clock would run fast when warm and would run slow when cold. He and his associates proved temperature-independence or temperature-compensation of the clock of *Uca* in their studies of circadian cycles of color change in fiddlers, again using specimens collected in the vicinity of Woods Hole. The black and the white color cells or chromatophores whose degree of pigment dispersion or concentration determines the gross coloration of the legs of the crabs were "staged" at 1:00 AM and 7:00 AM and at 1:00 PM and 7:00 PM. The staging method of Hogben and Slome (1931) was used, as it has been in many later observations of color changes in animals. Stage one of that system describes pigment in the punctate, most concentrated, condition; stage three refers to the intermediate degree of dispersion and stage five is assigned to a cell in which the pigment is maximally dispersed. Stage two is intermediate to stages one and three, while stage four lies between stages three and five (Fig. 3).

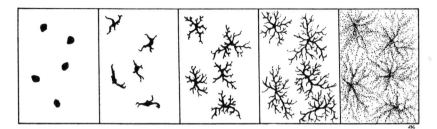

Figure 3.   Melanophores in Stages 1 (left) through 5 (right).

For their first series of experiments on the effects of temperature on the color change rhythm, Brown and his students kept three groups of *Uca* in constant darkness in buckets with a small amount of sea water. One of the groups was maintained at plus 26° C, the second at plus 16° C and the third at plus 6° C. The periods of the rhythms of changes in the dispersion of both the white and the black pigments were the same in all three groups of crabs — 24 hours. Only the amplitudes of the cycles varied, with that for the *Uca* at 16° being lesser than that for the animals at the high temperature, but greater than that for the crabs at the low temperature (Fig. 4).

A second series of experiments proved that the regulator of color change was, in part, metabolic. Crabs were immersed in sea water between zero and plus 3° C for 6 hours. After that treatment, they and their controls, which had been at a higher temperature, were placed in the dark at 16° C. Again, the average chromatophoric indices were determined at four times of day: 1:00 AM and 7:00 AM, 1:00 PM and 7:00 PM. The phases of the cycles of the chilled crabs were approximately six hours late compared with those of their controls (Fig. 5). They had been delayed by the cold. The new phasing, apparently set as the crabs warmed up, was retained under the constant laboratory conditions for as long as the crabs were observed. There was no tendency for the daily rhythms of the experimental *Uca* to return to the phasing typical of the control animals. That finding paralleled the one that Megušar reported for color changes which had been set by reversed light-dark treatment. Phases of the crabs' rhythms are not bound to specific events of the geophysical solar

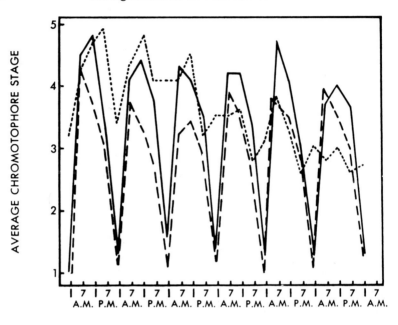

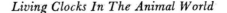

TIME OF DAY

Figure 4. Daily variations in the average stages of the melanophores of *Uca* maintained in constant darkness at 26° C (solid line), 16° C (dashed line) and 6° C (dotted line). From: Brown, F.A., Jr. and Webb, H.M.: Temperature relations of an endogenous daily rhythmicity in the fiddler crab, *Uca. Physiol Zool, 21*: 371-381, 1948. Copyright 1948 by University of Chicago Press.

cycle. They can be set at various real times by appropriate Zeitgebers.

Chilling fiddler crabs during different quarters of the day does not have identical effects on the phasing of their rhythms of color change. We had groups of *Uca* at 5° C from midnight to 6:00 AM, from 6:00 AM until noon, from noon until 6:00 PM or from 6:00 PM until midnight. In addition, we exposed other groups to the low temperature during different two-hour periods of the solar-day. In all cases, a delay in the phasing of the circadian rhythm was observed, but the degree of the delay varied with the specific solar times of chilling. Low temperature in the morning evoked the greatest delay; low temperature in the even-

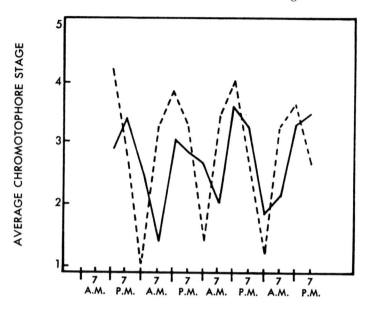

Figure 5.   Daily variations in the average stages of the melanophores of control (dashed line) and of experimental *Uca* (solid line) maintained in constant darkness. Before the staging of chromatophores was begun, the experimental crabs were exposed to 0° to 3° C for 6 hours. From: Brown, F.A., Jr. and Webb, H.M.: Temperature relations of an endogenous daily rhythmicity in the fiddler crab, *Uca. Physiol Zool, 21*: 371-381, 1948. Copyright 1948 by University of Chicago Press.

ing caused the least delay. Stephens (1957) also found varying sensitivity to changes in temperature in the fiddlers, and was able to correlate the amount of delay in the phasing of color change with the number of exposures to the lower temperatures.

Comparable "rhythms of sensitivity" to changes in light intensity have been known for *Uca* for more than twenty years (Webb, 1950). Again, it was the daily cycle of color change and shifts in its phases that indicated variations in the efficacy of dark-light and light-dark perturbations. Six-hour advances of the phases, which persist after the crabs are placed in complete darkness, can be induced by exposing the animals to bright light for durations of 6 or 12 hours if the 18- or 12-hour period of darkness, *i.e.*, the remainder of one solar period, ends at 1:00 PM. Light periods of 16, 20 or 24 hours with dark periods of 8, 4 or 0 hours, respectively, delay by 6 hours the phases of the fiddler crabs' pigmentary rhythm if the light period ends at 7:00 AM.

As Megusar had observed, Webb found that darkness during real day and bright light during the night caused a reversal or a 12-hour shift in the phases of the cycle. Once more, we see emphasized clearly the flexibility of the phasing of circadian cycles. However, 16 hours of illumination followed by 16 hours of darkness did not entrain the rhythm even after nine successive 32-hour blocks of light-dark treatment were employed. Therefore, circadian (*about* a day or 24 hours) frequency of an organismic cycle was clearly established. The color change rhythm could not be entrained to 32 hours. Examples to be cited in later chapters prove that daily cycles of other animals can not be made to repeat with periods which diverge very much from 24 hours.

But, how exact is the frequency of the rhythm of color change in fiddler crabs? Is the period of that cycle precisely 24.0 hours, or is it only approximately 24.0 hours? Often, it is stated that the daily rhythm of color change in *Uca* is one of the most exactly repeated solar cycles known, for it goes on day after day, for several months in complete darkness, without losing or gaining time. However, in many investigations, the staging of the chromatophores or the measurement of the indicator process has been done at most 24 times per day, often "every hour on the hour." The schedules of earlier investigations of that rhythm called for readings of the extent of pigment dispersion only four or six times in each 24-hour period. The variations in the free-running periods of individuals or groups of many other species are often measured in terms of minutes.

Securing average chromatophoric indices for groups of crabs every hour might well cover up such small variations; and, indeed, we do have evidence that the frequency of *Uca's* cycle of color change is merely *circadian*. The free-running periods of normal individuals' rhythms range between 23.5 and 24.3 hours. Under constant low illumination, even the periods of mean cycles for groups of fiddlers are longer than 24.0 hours. That apparent lengthening of the period when the animals are maintained in a lighted laboratory contradicts or provides an exception to Aschoff's Rule, which states that light active animals (fiddler crabs are diurnally active in their natural habitats) have a shorter circadian period in constant light than in constant darkness. (The converse is held to be true for night active species.) Since the precision of the color change rhythm continues to be debated, more detailed analyses of that cycle should be attempted. Studies of the influence of members of a group of crabs on the rhythms of one another should also be extended.

While working in Brown's laboratory at Woods Hole, several of us discovered that we had altered the color change cycles of fiddler crabs by keeping individuals isolated in separate containers. Under that condition, the chromatophores exhibited little or no daily variation. Hence, the curve of the cycle based on the average of the readings of color cells of the isolates was virtually a straight line, since its amplitude was so low. Yet only 8 hours after 18 of the crabs were combined in a common container, changes in the pigments became obvious, and an average rhythm of normal or usual amplitude was established. Important interactions among animals and the necessity of knowing the details of the rhythms of individuals in relationship to that of the group were suggested by our preliminary findings and by observations of some rhythms of honey bees, to be described in Chapter 4.

Stephens (1962) compared the cycles of fiddler crabs, some of which were isolated and others of which lived in groups. His findings allowed partial clarification of the contributions of different individuals to the rhythm of a population, especially in terms of frequency. For as long as 63 days after having been placed in constant darkness, the average rhythm of a group of *Uca pugnax*, known individually but living together, had periods varying between 22 and 26 hours. The same variation was true for the average cycles of groups of crabs, the members of which had been isolated from one another during the 63 days of darkness.

The periods of the rhythms of the individuals, whether they had lived in a group or in separate containers, also only approximated 24.0 hours. The range of their periods was not reported, but differences from 24.0 hours were extensive enough "to produce cycles completely out of phase with the day-night cycle at some point in the observations" (p. 932) in about one-half of the isolates and in one-third of the crabs living in common containers. No evidence of group entrainment was found. Thus, more evidence that the often reported precision of the frequency of the fiddler crabs' color change is a group rather than an individual phenomenon was adduced by Stephens.

The latter point was reinforced by computer analyses of the frequencies of the individual cycles of sham-operated and blinded *Uca pugnax* which lived under constant low illumination (Stephens, Halberg and Stephens, 1964). During 40 days, both groups maintained average cycles of 24 hours. However, the periods of the rhythm of the individual blinded crabs varied from 22.7 to 24.3 hours, and those of their controls, the sham-operated animals, ranged between 23.5 and 24.3 hours.

No matter which method is used, the determination of the exact periods of the daily cycles of color change in *Uca pugnax* will be difficult. That difficulty will be extreme during the first two or three weeks of maintenance of fiddler crabs in complete darkness. During those weeks, phases of the circadian rhythm are warped or skewed relative to solar time by a second persistent cycle, one of primary tidal frequency. That fascinating rhythm was discovered by Brown and his associates in 1952 (Brown, Fingerman, Sandeen and Webb, 1953). The 12.4-hour cycle of dispersion of the pigments is superimposed on the circadian one which, as described by Megušar, is simple in form. Generally, the legs of the crabs are darkest about midday and lightest about midnight with intermediate hues during mornings, afternoons and evenings. The persistent tidal effect, most obvious for the first 20 to 25 days after the fiddlers have been placed in the dark when the amplitude of the daily cycle is not yet maximal, tends to increase dispersion of the pigments from one to three hours after the times of low tides on the animals' native beaches (Fig. 6). The curves in Figure 6 illustrate those temporal relationships for six days in the summer of 1952 on which low tides occurred at very different hours. The indications of the tidal cycle were most often minor increases in dispersion of pigments — "notches"

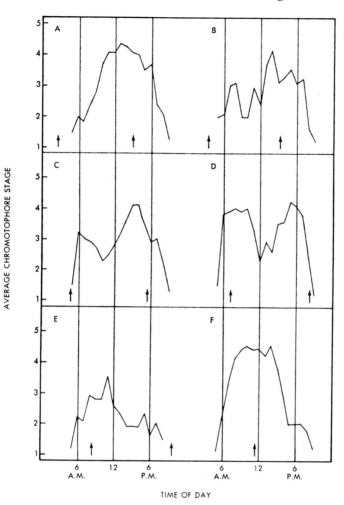

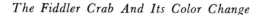

Figure 6. Daily variations in the melanophores of *Uca* maintained in constant darkness, and determined on the days immediately following the crabs' collections. The arrows indicate the times of low tides on the animals' native beach. A. August 6, 1952; B. July 24, 1952; C. August 9, 1952; D. August 12, 1952; E. July 31, 1952; F. August 9, 1952. From: Brown, F.A., Jr., Fingerman, M., Sandeen, M.I. and Webb, H.M.: Persistent diurnal and tidal rhythms of color change in the fiddler crab, *Uca pugnax. J Exp Zool, 123:* 29-60, 1953.

— which interrupted otherwise smooth falls and rises of the curves.

The timing of the "notches" or the phasing of the 12.4-hour cycle was proved to be set by tidal events in the animals' original habitats by comparing the rhythms of two populations of crabs which lived side by side in the laboratory. One group had been collected from Chappaquoit Beach on the Buzzard's Bay side of Cape Cod, and the second had been brought to the laboratory from Lagoon Pond on Martha's Vineyard. Low tides at Lagoon Pond occur approximately four hours later than comparable ones on Chappaquoit Beach even though the two areas are but a few miles apart. The staging of the chromatophores of the two collections showed that the color change cycles of the Lagoon Pond crabs included the tidal dispersing effect four hours later than did the rhythms of the animals from Chappaquoit. The "notches" signaled the times of low tides on each group's home beach. To be discussed in Chapter 3 are the results of a study in which I compared the tidal cycles of locomotory activity of *Uca pugnax* from Lagoon Pond and Chappaquoit. That investigation proved that the phases of the rhythm of running are more closely associated with lunar than with actual tidal events. The details of persistent cycles, even those of one particular species, vary to a surprisingly great degree.

The primary importance of local tidal events in the setting of phases of the color change rhythms of fiddler crabs was also emphasized by the observations of Fingerman and his students (1960) on *Uca pugilator* and *Uca speciosa* from beaches of Louisiana and Mississippi. They had rhythms of color change similar to those of *Uca pugnax* from Massachusetts. Tidal "notches," seen about 50 minutes later on succeeding solar-days, were superimposed on very obvious daily rhythms of darkening and lightening. Under constant laboratory conditions, solar and tidal cycles persisted. However, the tidal dispersions of pigments of *Uca speciosa* occurred almost five hours in advance of those of *Uca pugilator*. That surprised the investigators, because all the crabs had been collected from the same beach. What phase-setting perturbations or environmental factors varied by five hours for the two different species in the same general habitat?

The times at which their burrows were uncovered by the

sea varied, for *Uca speciosa* lived higher on the beach than did *Uca pugilator*. Consequently, following high tides, the burrows of *Uca speciosa* were uncovered by the receding waters earlier than those of the second species. The crabs on the upper reaches of the shore experienced low tides in advance of the animals that burrowed closer to the ocean. Fingerman was able to prove for those two groups of fiddlers and for two populations of *Uca pugilator*, which burrowed at different levels of the beach, that the phases of the tidal rhythm of color change were set to the times when the animals' burrows were uncovered by the ebbing tides and at which times the crabs began to emerge to run on the beach.

Other *Uca pugilator* which lived above the high tide line near Ocean Springs, Mississippi, exhibited no tidal cycle of pigment dispersion, although under laboratory conditions, they had a typical daily rhythm of color change. Their burrows are probably covered by tides only during severe storms, and they undoubtedly are not forced to live in time with the tides. If they migrated to the intertidal zone, and ran and fed there only during low water, how long would it be before a tidal cycle of color change were seen? What are the Zeitgebers or entraining factors of *Uca*'s tidal rhythm that set its phases in the crabs' natural environments?

Under laboratory conditions, some light-dark changes set the phases of the 12.4-hour cycle of color change just as they set the phases of the circadian one. Since the directions and the degrees of the shifts of events evoked by L/D regimens were the same in the daily and the tidal rhythms, it was concluded that the two cycles had maintained their usual temporal relationships with one another (Brown, *et al*, 1953). Since that is true, one might assume that sojourns at temperatures below 5 to 6° C, which delay the phases of the daily cycle of darkening and lightening, would also shift the tidal "notches" in the same direction, *i.e.*, late in time, and by the same number of hours as events of the solar cycle were delayed. We do know that the frequency of the 12.4-hour rhythm is independent of temperature between 13 and 30° C, although the amplitude of the cycle is depressed when the crabs experience temperatures in the range of 13 through 22° C (Brown, Webb, Bennett and Sandeen, 1954).

The temporal relationships between the solar and tidal rhythms of color change in *Uca* can account for a third persistent cycle, one of a semilunar or approximately 14.8-day period. Since the supplementary dispersions of pigments, the manifestations of the tidal cycle, move later relative to the daily cycle by an average 50 to 52 minutes per solar-day, it is only every half-month that the phases of the two rhythms lie in identical relationship with one another. Semilunar cycles were found for *Uca pugnax* by Brown's group, and for *Uca pugilator, Uca minax, Uca speciosa* and the blue crab, *Callinectes sapidus,* by Fingerman and his students.

An especially clear demonstration that fiddler crabs' pigmentary responses vary at semilunar intervals was provided by Webb (1966). She was able to correlate the degree of dispersion of the black pigment of *Uca pugnax* at 6:00 PM with the phase of the moon. On the few days preceding new and full moons, the average chromatophoric index of groups of ten crabs was low, between 1 and 2. By four days after new and full moons, the averages for the same groups were high, above 4. The same correlations held for crabs exposed to natural day-night changes in illumination from July through December, and were, therefore independent of photoperiod.

The existence of semilunar cycles in fiddler crabs is certainly clear. What is not clear is whether or not the animals use them in the timing of any of their activities. Their reproductive behavior is not synchronized with either phases of the moon or the tides, as is that of the Atlantic fireworm or the Palolo worm. The semilunar high, high and low, low tides would seem to have only minor effects on the lengths of the crabs' feeding periods. Possibly, the fiddlers by virtue of their having both solar and lunar rhythms are able to measure off the passage of a year. The geophysical cycles of the sun and the moon come into identical relationships but once annually. Is that also true of organismic cycles of solar and lunar frequencies? It would seem adaptive for *Uca,* especially those of the temperate zones of the world, to be able to function at levels which vary appropriately with the changing seasons of the year.

An animal, equipped with cellular clocks which beat out inexorably the day, the tides, the months and the year, and that have hands or indicators which are labile enough to be set or reset by naturally occurring Zeitgebers, would enjoy the great advantage of fitting precisely in its rhythmically changing surroundings. Our knowledge of color change, one indicator of the clock of fiddler crabs, supports the argument that these crustaceans do enjoy that advantage.

The effects of some temperature changes on the hands of living clocks, their phases and the amplitudes of their cycles, also illustrate the great adaptive flexibility of biological timing mechanisms. The temperature-independence or temperature-compensation of persistent rhythms, including the circadian and tidal ones of *Uca,* certainly contributes to the remarkable stability of the cellular clockworks. We proved the temperature-independence of the tidal cycle of fiddler crabs using a technique worked out by Hines (1954). Hourly through the day, she staged black chromatophores of legs of *Uca pugnax* which had been isolated in small dishes of sea water following their autotomy or self-amputation. That casting off of appendanges can be evoked by pinching, just slightly, the distal ends of the legs. In those isolated legs, the degree of dispersion of the pigment immediately after and 30 minutes after autotomy reflected the time of day and the phase of the tide on the animals' native beaches.

Compared with the index of the chromatophores at the moment after isolation of the legs, the index, or measurement of the degree of dispersion, was lower one-half hour later. Some concentration of the pigment always occurred. However, less concentration was found in tests run between 8:00 AM and 2:00 PM than between 2:00 PM and 7:00 PM. Within that 8:00 AM to 7:00 PM span of the solar-day, greater pigment concentration was seen near the times of low tides and lesser concentration around the times of high tides at Chappaquoit, the beach from which the *Uca* were collected. Whether we kept our crabs at 13° C, 22° C or 30° C, the frequency and the phasing of the tidal cycle of color change in the isolated legs were the same.

Changes in the pigmentation of autotomized legs of *Uca pugilator* taught us that even crabs which have lost their eyestalks have a diurnal rhythm of color change (Webb, Bennett and Brown, 1954). Abramowitz (1937) was convinced that the rhythm disappeared after crabs were destalked. Sources of pigment-dispersing hormones which mediate the cycle are removed with the eyestalks. Nevertheless, we found that the black pigment in legs isolated from animals which had been destalked 8 to 48 hours before autotomy, and which was at stage one immediately after autotomy, dispersed within 30 minutes when we ran tests between 2:00 AM and 8:00 PM. From 8:00 PM through 2:00 AM, the dark pigment did not show any change from stage one. Even in those highly abnormal preparations, we had evidence of a persistent daily rhythm of color change characterized by darkening during the day and lightening during the night.

Our observations of the changing pigmentation in the autotomized legs were supported by the results of a series of injection experiments with destalked *Uca pugilator*. During 24-hour periods, we treated those crabs with extracts of eyestalks of the same species. The extracts caused less dispersion at night than the same doses and concentrations caused during the day. In other words, the chromatophores varied in their reactions to the color change hormones contained in the extracts. How are we able to explain that difference? How can we explain the cycle of color change in eyestalkless fiddler crabs?

The explanations rest on two major points which were emphasized by our work and by many other studies of the color change hormones or chromatophorotropins of crustaceans. (See reviews by Fingerman, 1970a and 1970b). 1) The neurosecretory complex of the eyestalk is not the only source of color change hormones. Additional parts of the nervous systems of crabs and their relatives contain cells which also produce chromatophorotropins. 2) There are several different hormones: a black-pigment dispersing hormone, a black-pigment concentrating hormone, as well as ones which individually cause dispersion or concentration of the white or red pigments of crustaceans.

Our studies of the stalkless *Uca pugilator* underscored the antagonistic actions of black-concentrating and black-dispersing chromatophorotropins. We postulated that during the night, there was a greater titer of a black-concentrating hormone of noneye-stalk origin in the body fluids than there was during the daytime. Therefore, the black-dispersing hormone of the extracts which we injected could have an obvious effect only during the day when the amounts of its antagonistic hormone were low. The black pigments in isolated legs of destalked animals disperse more diurnally than nocturnally, because during daylight hours there is less black-concentrating hormone in the blood than there is during the remainder of a circadian period. According to the extensive investigations of Fingerman and his students, the black-pigment concentrating chromatophorotropin may be produced in the postcommissure organ which is located in the heads of crabs. Amputation of the eyestalks would not affect the supply of that factor.

The color change rhythm of calico-backs may also help us secure more detailed information about two other chromatophorotropins, a red-concentrating factor and a red-dispersing factor. Intact (Brown, 1950) and eyestalkless (Fingerman, 1968) crabs kept under constant conditions of darkness or low light intensity have persistent circadian cycles of migration of the red pigments in particular chromatophores. The pigment disperses by day and concentrates by night as does the black pigment. However, the phases and the amplitudes of the solar cycles of the changes in the red and black pigments are not the same. Different hormones are undoubtedly involved in the mediation of those two indicators of circadian change in *Uca pugilator*.

Do the neurosecretory cells which synthesize and secrete the various factors respond differently to common temporal cues coming from the crabs' biological clocks? Are the several different chromatophorotropins effective for varying periods of time? Is it possible that the chromatophores, themselves, have latent periods to different hormones which are not alike? When one attacks the many problems of the endocrinology of color change in *Uca*, the details of their persistent daily and tidal cycles of movements of pigments should be kept in mind. Only observations and experiments which are temporally comparable can lead to conclusions which have clear and valid meanings.

In some of our approaches to questions of the influence of geophysical factors, other than light and temperature, on persistent cycles, we have used the color change rhythm of *Uca pugnax* as our assay. We exposed groups of that species to showers of particles of cosmic-ray origin whose intensities were increased approximately 50 per cent by lead plates placed at proper distances over the crabs (Brown, Bennett and Ralph, 1955). Control crabs were exposed to showers of normal intensity. Often, control and experimental animals were exchanged with each other to reduce the possibility that differences between them were attributable merely to minor color variations existing between any two populations of crabs.

To obviate any possible bias on the parts of the investigators, assistants gave the crabs to the person who was staging the chromatophores without informing that individual whether the crabs had been under the lead plates or under control conditions. Those experiments spanned a three-month period during which time more than 16,000 observations of individual crabs were made. Did the change in the intensity of cosmic radiation influence the rhythm of pigment dispersion and concentration? Yes, and the fiddlers' response, itself, was rhythmic in character. During most of the day, the chromatophoric indices of the crabs under lead were consistently higher than those of the control animals (Fig. 7). However, in the early morning hours, 1:00 AM to 4:00 AM, the opposite situation obtained; the black pigment of the experimental *Uca* was less dispersed than that of the controls. The differences between the groups, those under lead and those not, were small, but they were reproducible and were statistically significant.

Of what significance are these findings in terms of understanding living clocks? A subtle geophysical factor, fluctuating within the range of magnitudes which actually exists on earth, affected the amplitude of a circadian cycle. Were its effects on the works of the clock, the mediating pathways of the clock or on the hands of the clock? Whatever the case, the direction of its effects varied during the solar-day. That gave us more evidence of rhythmic variations in the sensitivity or reactivity of the living

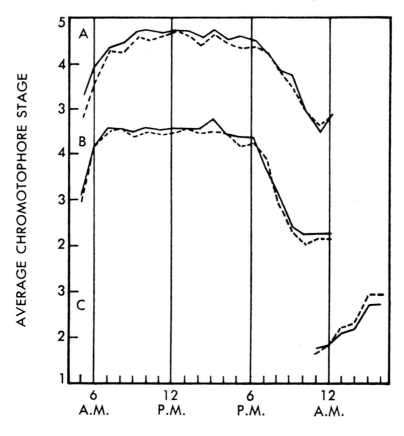

TIME OF DAY

Figure 7. Variations in the average stages of the melanophores of experimental *Uca* (solid line) and of their controls (broken line). The experimental animals were kept under lead while the controls were under wood. A. May 25 – June 6, 1954; *B.* June 18 – 21, 1954; *C.* July 19 – 21, 1954. From: Brown, F.A., Jr., Bennett, M.F. and Ralph, C.L.: Apparent reversible influence of cosmic-ray-induced showers upon a biological system. *Proc Soc Exp Biol Med,* 89: 332-337, 1955.

systems, themselves. What might be the result of changing the times, in terms of the crabs' sensitive periods, at which the animals are exposed to specific phases of the cycles of geophysical factors such as cosmic radiation? Do living clocks require particular environmental stimuli at specific phases of their own cycles of sensitivity in order to continue their precise timekeeping? The answer to the latter question is no; and it, too, comes to us in part from a study of color change in *Uca pugnax* (Brown, Webb and Bennett, 1955).

The fundamental aim of that study, its protocol and its results were very similar to those of an investigation focused on the cyclic behavior of honey bees. The work with the bees was planned in 1937 by von Frisch, and was conducted in 1955, a year after our comparable experiment, by Renner. The von Frisch-Renner study is discussed in detail in Chapter 4. The questions to be answered by those investigations were these: can animals continue to mark off time precisely if the phases of their environmental rhythms are changed in time relative to the organisms' physiological cycles? How necessary are cyclic exogenous factors to the maintenance of the frequency or period of persistent rhythms? Are biological timing mechanisms driven or even modulated by regularly recurring external events?

The investigations were designed on the basis of these points: one can not shield his organisms from or compensate for all environmental changes which depend upon the relative positions of the sun, the moon and the earth. Therefore, at no point on the surface of the earth can an investigator modify completely the normal temporal relationships which exist between living systems and their surroundings. However, given rapid enough transportation, he can effect such modifications by moving his organisms from their original time zone to another where environmental changes occur earlier or later than in the original habitat and, therefore, at different phases of the organismic rhythms. For his first experiment with honey bees, Renner transported his animals between Paris and New York. For his second study he was, as were we, less international, and flew his bees between the east and west coasts of the United States.

Similar rooms in the Marine Biological Laboratory in Woods Hole, Massachusetts, and in the Department of Zoology, the University of California in Berkeley, housed our experiments. Woods Hole and Berkeley are separated by 51° longitude, and solar and lunar events occur three hours later in California than on Cape Cod. We collected all our crabs on one afternoon from Chappaquoit Beach near Woods Hole. Half of that collection was kept in the dark in our laboratory; the other half was taken to Berkeley in a covered bucket by Professor Webb via commercial airlines. During the day of transit (12 hours of travel), the animals could not have received any external signal in its normal 12.4- or 24.0-hour frequency, and once in California, such signals would have come three hours later in the crabs' rhythms than they had in Massachusetts.

At the same moment, 10:00 AM, P.S.T. and 1:00 PM, E.S.T., crabs were removed from their darkened containers and placed in sea water where they were exposed to constant low illumination. The staging of the chromatophores commenced, and continued in the two laboratories at the same hours for the next seven days. Figure 8 presents the data for the two groups of crabs during that period of observation. All the average chromatophoric indices have been plotted against Eastern Standard Time. The arrows indicate the times of low tides on Chappaquoit Beach. Even on the first day of study, the curves representing pigment dispersion were at most only a few minutes apart, and through the next six days, there was no constant tendency for the phases of the rhythms to drift away from one another. The *Uca* in California had remained on Massachusetts time. Although their cycles were in different relationships with the rhythms of most geophysical factors, the crabs on the Pacific coast had been able to mark off accurate solar and lunar periods for a week's time. The endogenous component of the clock of the fiddler crab was proved to be a stable one.

As has been emphasized, a living clock which is stable enough to maintain its frequency in the face of many perturbations, but whose mediating pathways and hands are flexible enough to be synchronized with events of the physical world that are of importance to organisms, is highly adaptive to its possessor. But, are the oscillations or changing levels of indicator processes of direct benefit to their possessors? Does the color change rhythm of *Uca*, which is most obvious on the surfaces of the animals' legs, lend any distinct advantages to fiddler crabs? Most fiddlers run

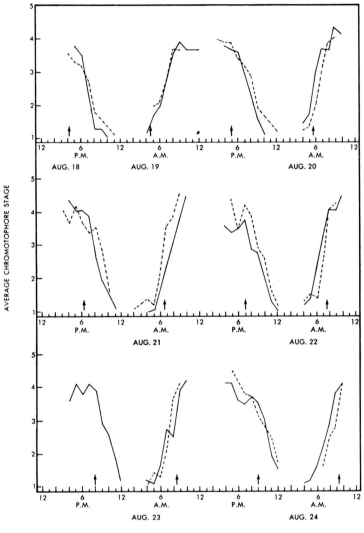

Figure 8. Variations in the average stages of the melanophores of *Uca* in California (broken line) and in Massachusetts (solid line). The arrows indicate the times of low tides on the crabs' native beach. From: Brown, F. A., Jr., Webb, H.M. and Bennett, M.F.: Proof for an endogenous component in persistent solar and lunar rhythmicity in organisms. *Proc Nat Acad Sci, 41*: 93-100, 1955.

on their beaches during low tides of the daytime. Most of their pigments are more dispersed during the day than during the night. Furthermore, these same pigments tend to be more dispersed around the hours of low tides than around the times of high tides. Is it truly adaptive to the crabs that their chromatophores are in the dispersed state when the animals are out of their burrows, running, waving and feeding?

Pigmentation and its changes may function in several important manners in the lives of the fiddlers. There is evidence of thermoregulatory roles (Fingerman, 1970b). Barnwell's comparative studies (1963 and 1968) of equatorial, tropical and temperate zone species provide us with some information about coloration, camouflage and protection of *Uca*. However, those studies as well as many by Crane (1944 and 1958) on the behavior of various fiddler crabs caution one against making general statements. The physiological, behavioral and ecological characteristics of each species of *Uca* must be considered in all combinations when one attempts to work out the adaptive nature of cycles of color change to fiddler crabs.

Chapter 3

CRABS, THEIR RHYTHMS OF METABOLISM AND ACTIVITY

An investigator asking questions of the clock of fiddler crabs is not limited to seeking answers from their cycles of color change. In his 1954 article in *Scientific American*, Frank Brown was able to tell his readers about another indicator of *Uca's* timing mechanism, for during the previous summer, we had found persistent solar and lunar rhythms of oxygen-consumption in both black-backs, *Uca pugnax*, and in calico-backs, *Uca pugilator* (Brown, Bennett and Webb, 1954). To secure those data, it had not been necessary for investigators to make measurements throughout entire 24-hour periods, as must be done when the

44

stages of pigment dispersion in chromatophores constitute the parameters of cycles. Records of the levels of the crabs' consumption of oxygen were obtained continuously and automatically.

A photograph of the first model of the automatic, continuous recording respirometer which had been built in Brown's laboratory and used for our studies of the fiddler crabs was included in *Scientific American.* Later models which allowed for the organisms whose metabolic fluctuations were being monitored to be kept under constant barometric pressure, *i.e.*, to be sealed away from the atmosphere, have also been described and diagrammed (Brown, 1960). The vessels of the respirometers which contain the plants or animals and the chemicals which absorb gaseous wastes are connected to a plastic bag of oxygen by a hypodermic needle (Fig. 9*a*). The entire unit, vessel and bag, functions as a diver which is suspended in a constant temperature bath and is attached to the arm of a sensitive ink-writing lever (Fig. 9*b*).

As oxygen is used by the organism, more of the gas flows from the bag into the vessel. Consequently, the unit becomes less buoyant. The changes in buoyancy, reflected by changes in weight, are recorded on a slow moving kymograph. Measurements of the oxygen-consumption of undisturbed animals or plants can be made over three to four days' duration. Workers in Brown's laboratories and in others have successfully used the continuous recording instrument in their investigations of respiration in fiddler crabs, insects, earthworms, salamanders, seaweeds, potatoes and carrots and for studies of variations in oxygen tension during photosynthesis.

Figure 10 illustrates the average solar cycle of oxygen-consumption for *Uca pugnax* during a block of 15 consecutive days in the summer of 1953. The crabs had been collected from the mud flats at Chappaquoit on Buzzard's Bay near Woods Hole. During the recording of their oxygen-consumption, the animals lived in their respirometer flasks at a constant temperature and under constant low illumination. The records show an obvious peak of the respiratory rate around 7:00 AM. A period characterized by much lower use of oxygen extends from just before

A.

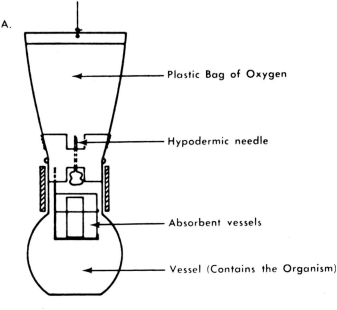

Plastic Bag of Oxygen

Hypodermic needle

Absorbent vessels

Vessel (Contains the Organism)

RESPIROMETER

Figure 9. *A.* Respirometer with the vessel for an organism connected to the plastic bag of oxygen by a hypodermic needle. *B.* (p. 47) Respirometers in baths, hermetically sealed. The ink-writing lever and the kymograph are also shown. From: Brown, F.A., Jr.: Response to pervasive geophysical factors and the biological clock problem. *Cold Spring Harbor Symp Quant Biol,* 25: 57-71, 1960. Copyright 1961 by Cold Spring Harbor Laboratory.

midday until 7:00 PM. A very similar circadian cycle of metabolism was found for *Uca pugilator* during the same 15-day period.

Frequently, mean or average solar cycles for 15- or 30-day blocks of time have been calculated and presented as examples of rhythms of oxygen-consumption. Why are hourly data from 15 consecutive days used? The rationale is this: if there is a tidal influence on the rate of metabolism or on that of any other indicator process, the effect can be expected to have influenced the process at all hours of the solar-day during roughly 15 days or a semilunar period. That is true at least in areas where within each day there are two high tides and two low tides. Each tidal event moves to later times of the solar-day by an average 52 minutes

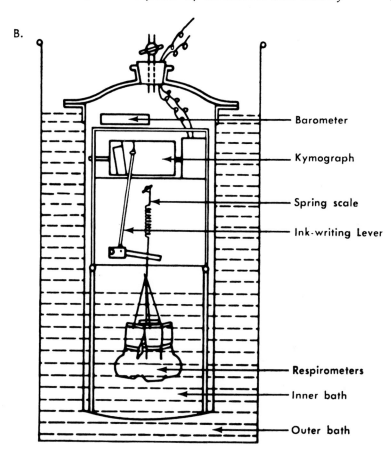

RESPIROMETERS IN BATHS

per day, and at the end of every 14.75-day period, each hour of the solar-day will have experienced one high tide and one low tide. Therefore, the data for every hour of an average 15-day circadian cycle include the possible influence of comparable tidal phases (Fig. 11).

Likewise, 30-day averages contain within their mean hourly figures the possible effects of lunar zenith and of lunar nadir, for those phases are separated from each other by 12.4 hours, and also move across the solar-day at an approximate rate of 52 minutes per day. Hence, a particular lunar phase recurs at the same solar time only every 29.75 days (Fig. 12).

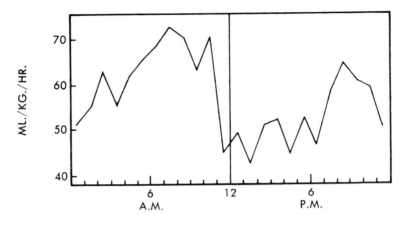

TIME OF DAY

Figure 10. The average 15-day solar cycle of oxygen-consumption of *Uca pugnax*, August 22 — September 5, 1953. From: Brown, F.A., Jr., Bennett, M.F. and Webb, H.M.: Persistent daily and tidal rhythms of O₂-consumption in fiddler crabs. *J Cell Comp Physiol, 44*: 477-505, 1954.

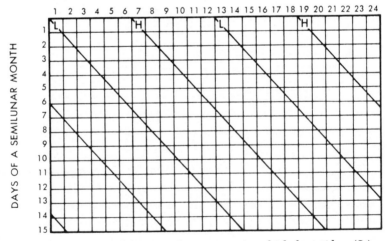

Figure 11. A diagram showing the manner in which low tides (L) and high tides (H) scan the hours of solar-days through a 15-day or semilunar period. The tidal events move to later solar times by an average 52 minutes per day.

HOURS OF SOLAR DAYS

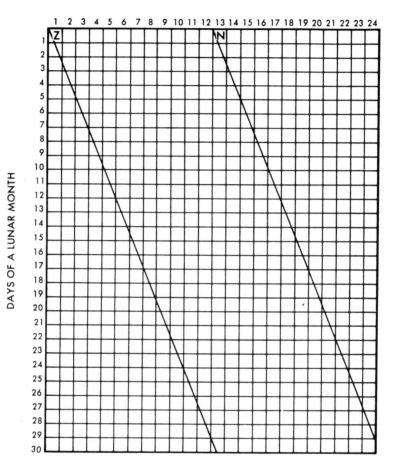

Figure 12. A diagram showing the manner in which lunar zenith (Z) and lunar nadir (N) scan the hours of solar-days as they move to later times by an average 52 minutes per day.

To sharpen the forms of tidal or lunar rhythms of oxygen-consumption or of any physiological activity, the hourly data from 15 or 30 consecutive days can be treated in terms of tidal or lunar "hours." The method used is a simple one, originated by Laplace, an eighteenth century mathematician and astronomer, during his investigations of tides of the atmosphere. The late Sir

Sidney Chapman also adopted this scheme for his analyses which established the reality of lunar effects on the rhythmic movements of the great air masses surrounding our earth. Essentially, one averages data not for the same hour of successive solar-days, but those for hour 1 of day n, with those for hour 2 of day n plus 1, with those for hour 3 of day n plus 2, *ad finem* (Fig. 13). Similarly, the value for hour 2 on day n, that for hour 3 on day n

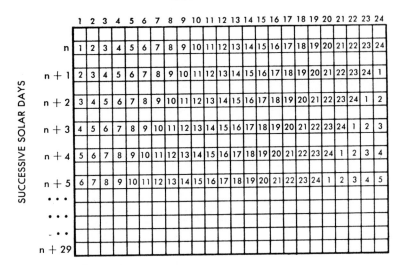

Figure 13. The method of Laplace for the analysis of tidal or lunar influences. The hourly data are moved one hour early to line up figures which include comparable tidal or lunar effects.

plus 1, that for hour 4 on day n plus 2, etc. are averaged. In other words, one slides the hourly data of each succeeding day earlier by one hour each day. Since tidal or lunar phases actually move to later times by roughly 52 minutes per day, their possible effects are lined up vertically by moving hourly data earlier by an hour per solar period.

If one uses 15 days of data, appropriate when tidal influences are indicated, the tidal effects exerted during hours 1 through 12 of solar-days are aligned, and then averaged, while those of hours 12 through 24 are aligned and averaged. When one incorporates

the hourly data of 30 days in the Laplace scheme, he lines up
the figures influenced by lunar zenith in one column and those
affected by lunar nadir in another column, that representing a
time about 12 to 13 hours later.

In Figure 14 is presented an average 15-day tidal rhythm of

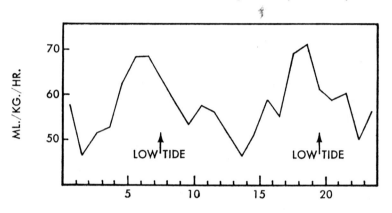

Figure 14.   The average 15-day tidal cycle of oxygen-consumption of
Uca pugnax, August 22 — September 5, 1953. The arrows indicate the times
of low tides on the crabs' native beach. From: Brown, F.A., Jr., Bennett, M.
F. and Webb, H.M.: Persistent daily and tidal rhythms of O₂-consumption
in fiddler crabs. *J Cell Comp Physiol*, 44: 477-505, 1954.

oxygen-consumption for fiddler crabs, *Uca pugnax*, recorded during our initial study of metabolic cycles in 1953. The periods of
the greatest consumption of oxygen were just prior to the times
of low tides on Chappaquoit Beach from which the animals had
been collected. Those are the times when the supplementary
dispersion of pigments is seen, the times when the crabs are most
likely to be running and feeding on their flats (Chapter 2), and
the times at which crabs, kept under constant laboratory conditions, are most active. A description of *Uca's* activity cycles is
presented later in this chapter. A tidal cycle of oxygen-consumption, again very similar in general form, amplitude and phasing
to that of *Uca pugnax,* was found for *Uca pugilator,* which had
also lived at Chappaquoit.

The clear-cut methods used to sharpen such average tidal as

well as average lunar and solar rhythms were severely criticized in an often quoted, but erroneously conducted analysis by L. C. Cole (1957). He described cycles of oxygen-consumption for the unicorn which he claimed were generated by sliding his data, which were numbers from tables of random numbers, exactly as hourly data are moved when one uses the Laplace method of tidal or lunar analyses. However, Cole used only five successive days of hourly "data." Fifteen or 30, as discussed above, must be used. In addition, Cole moved his figures in the wrong direction, *i.e.*, to later rather than to earlier hours. Furthermore, as Heusner (1965) pointed out, Cole injected a nonrandom artifact in the manner in which he used his "data." The arithmetical techniques we have used to help describe average persistent cycles of oxygen-consumption and activity may be open to criticism and certainly can be improved, but Cole's points do not constitute valid criticism.

Another criticism which loses strength because of lack of precision in itself is that published by J. Enright (1965). Enright is well acquainted with cycles of tidal frequency, having described them for four different species of beach inhabiting crustaceans. He reported that he had reanalyzed by periodograms data for oxygen-consumption or activity of various organisms which had been recorded in the laboratories of other investigators. In all cases, it had been claimed that the organisms possessed both average solar and tidal or lunar rhythms. Of the 17 sets of data which he reanalyzed, Enright found evidence for such cycles in only three cases. He concluded that in the remaining 14 cases, the rhythms were artifacts of the Laplace method of analysis.

The periodogram is indeed valuable to biochronometry. But, its value is only so great as is the reliability of the data which it treats. Most of the original records of oxygen-consumption and levels of activity, the data Enright should have used in his periodograms, have never been published because of limitations of space or finances. He was able to use only average figures which were published or which he could read from published curves which described average solar and lunar-tidal cycles.

And, no matter how we analyze the variations — hour by

hour, day by day, month by month, fiddler crabs consume oxygen at varying levels, and those levels trace out cycles of solar, tidal and lunar frequencies. Furthermore, the phases of the crabs' persistent rhythms correlate adaptively with solar and tidal phases of the animals' natural environment, the intertidal zone of beaches and mud flats.

A series of reports (Webb and Brown, 1958, 1961 and 1965) provides us with details of the persistent rhythms of *Uca pugnax* as they were expressed in the laboratory during the summer months. Consistently, peaks of oxygen-consumption occurred at 12.4-hour intervals. Thus, again, there was overwhelming evidence that phases of a tidal or primary lunar cycle were scanning successive solar-days at an average rate close to 52 minutes per day. Also published (Webb and Brown, 1961) are the records of variations in *Uca*'s oxygen-consumption, measured as usual under constant laboratory conditions including barometric pressure, during the summer, the fall and part of the winter of one year. Differences in the average monthly levels of metabolism suggest an annual cycle of oxygen-consumption in the crabs, and those variations bear a direct and amazing relationship with the average monthly outdoor temperature in Woods Hole. The higher the average air temperature, the greater the average rate of oxygen-consumption. That correlation could not have indicated a direct cause-effect relationship between temperature and metabolism, for the crabs lived at 24° C from one to five months before recording and during the recording sessions themselves. Are there geophysical factors, fluctuating with an annual frequency, which affect the weather, as reflected by air temperatures, and affect the lives of the crabs as reflected by their rates of oxygen-consumption? Or was the relationship between temperatures and levels of metabolism merely fortuitous?

Questions about the mechanism of phase-setting of the lunar-tidal rhythms of oxygen-consumption in fiddler crabs have also been posed, and have been answered — at least, partially. *Uca pugnax* and *Uca pugilator* use tidally related cues to phase these cycles (Barnwell and Brown, 1963) as they do in setting the phases of their tidal cycles of color change (Chapter 2), and as they do in phasing their overt tidal rhythms of running activity which are described later in this chapter. Crabs from Chappaquoit Beach provided the information about the phasing of the metabolic cycles. The peaks of oxygen-consumption of *Uca pugnax* and *Uca pugilator* which had been collected from the upper areas of the beach were recorded from one to four hours earlier than comparable phases in the rhythms of animals whose burrows were located on the lower reaches of the tidal flats.

As was noted in Chapter 2, as the water recedes prior to the times of dead low tides, the crabs that live in the upper beach and farther from the low tide line, are uncovered earlier than the fiddlers which live closer to that line down on the lower areas of the beaches. The differences between the times of maximal oxygen-consumption for the two groups of crabs were seen to be related to the times at which they probably started to run and to feed in their original and individual habitats. In addition, *Uca pugnax* and *Uca pugilator* which lived closer to the sea had higher average rates of metabolism than did the specimens that burrowed farther from the ocean.

The forms of the rhythms of the crabs inhabiting the two parts of Chappaquoit Beach were the same as those of the *Uca pugnax* studied through many summers in Woods Hole. From year to year, the characteristics of the metabolic cycles, phasing and amplitude, as well as form, have been remarkably consistent. However, we recorded one exceptional case of a major and radical change in the solar cycle of *Uca pugnax* (Brown, Webb and Bennett, 1958).

The change occurred during the summer of 1954, and paralleled similar aberrations observed in several other organisms' rhythms which were well-known to us. These changes correlated in turn with one observed in the pattern of fluctuations in cosmic radiation recorded for the same blocks of time during which the organismic rhythms were atypical. In Figure 15, the relationships between the cycles of the levels of the nucleonic component of background radiation (very kindly made available to us by Professor J. A. Simpson of the Fermi Institute of the University of Chicago) and those of oxygen-consumption of fiddler crabs in

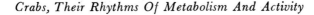

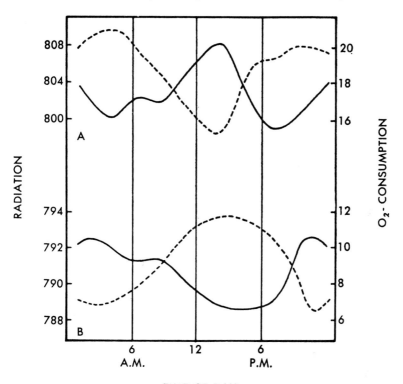

TIME OF DAY

Figure 15. A. The 3-hour moving mean of the average daily cycle of the nucleonic component of cosmic radiation for July 7 — August 4, 1954 (broken line) and the 3-hour moving mean of the average daily cycle of oxygen-consumption of *Uca pugnax* for the same period of time (solid line). B. The same as A., but for the period, July 6 — August 3, 1955. From: Brown, F.A., Jr., Webb, H.M. and Bennett, M.F.: Comparisons of some fluctuations in cosmic radiation and in organismic activity during 1954, 1955 and 1956. *Am J Physiol, 195:* 237-243, 1958.

two different summers, 1954 and 1955, are illustrated. The curves of radiation and metabolism are essentially inversions of one another. About the times of maximal oxygen-consumption, the intensity of radiation was low. Conversely, the hours of low average metabolism were the times of greater radiation. That basic relationship held both in 1954 and 1955, even though in 1954 the crabs' average daily rhythm was inverted relative to the one recorded in 1955, since the same type of inversion occurred between the cycles of radiation intensity.

As is discussed again in Chapter 6, there is no explanation available for the changes seen in the rhythms of the fiddler crabs, those of quahogs, those of potatoes or those of the nucleonic component of background radiation. Is it possible that other subtle geophysical factors exerted influences that could be read both in the measurements of the intensity of cosmic radiation and in the levels of the physiology of a number of organisms? A cause-effect relationship between the intensity of radiation and the workings of biological clocks can not be defended merely on the basis of the interesting and changing patterns which have been described.

A closely related problem is this: what is the meaning of the temporal relationships which exist between average persistent cycles of oxygen-consumption in animals and plants and mean rhythms of barometric pressure change? Do organisms and the earth's atmosphere respond commonly to variations in the levels of subtle geophysical factors? Our studies of *Uca pugnax, Uca pugilator*, salamanders (Chapter 7), seaweeds, potatoes and carrots have contributed circumstantial evidence that that may be true. The measurements of metabolism were corrected for concurrent changes in barometric pressure or were recorded while the organisms were in hermetically sealed respirometer vessels. The evidence was made even more secure by our finding that the opening and closing of the valves of clams and oysters (Chapter 6), the physical measurements of which could not have been affected by pressure changes, were also related to fluctuations in barometric pressure.

For the fiddler crabs, the situation was that their rates of respiration correlated negatively with the concurrent rates of changes in barometric pressure. The rate of oxygen-consumption rose as the barometer fell, and vice versa. Is that correlation, as well as the one to be described for the bivalve molluscs and barometric pressure and for salamanders and barometric pressure, only a coincidence?

Some authorities are of that persuasion. They find it probable that the correlations which have been found between persistent cycles of organismic activities and fluctuations of diverse geophysical factors are based on no true interrelationships. They are nothing but nonsense correlations. Others are not so convinced that only coincidentally do the two types of variations show statistical similarities. The correlations might indicate common generating or modulating phenomena, or even, in some cases, cause-effect relationships. How can we prove or disprove these points?

Most obviously, we can attempt to gain critical answers by sending our rhythmic organisms to extraterrestrial space, away from the influence of geophysical factors which cycle at earthly frequencies. What conclusions will we be able to draw if, after several weeks in space, the rhythms of metabolism of fiddler crabs or those of the running of cockroaches or those of the hatching of fruit flies die out or change in frequency? Conversely, what have we proved if, under conditions of outer space, the organisms' clocks and their indicator processes continue running as they did on earth under constant conditions of light, temperature, humidity and barometric pressure? When are we likely to have data from space?

Even if fiddler crabs are not sent to extraterrestrial laboratories, we have much more to learn about metabolic rhythms from them here on earth. Only two reports of entrainment and of phase-shifting of *Uca*'s cycles of oxygen-consumption are available. In both those cases, the animals were subjected to reversed illumination. They were in the dark by day and under light of approximately 10 lux by night. Before the crabs were placed in their respirometers, it was ascertained that their chromatophore rhythms had been reversed by the treatment. In both studies, the records of the variations in the levels of metabolism indicated that the diurnal cycles of the experimental animals had been modified. Essentially, the phases of the circadian metabolic rhythms had been reversed. However, the tidal cycles showed little if any change.

Whether or not the average circadian or tidal fluctuations of oxygen-consumption of fiddler crabs can be changed by temperature perturbations or by changes in the intensities of geophysical factors, *e.g.*, magnetic flux, radiation, is not known. That kind of information can be gained from investigations conducted on earth with equipment which is generally found in most laboratories.

More information valuable to our understanding of the biochronometry of fiddler crabs could also be gleaned from further analyses of their rhythmic locomotory activity. The cycles of running of *Uca pugnax* were first described by us in 1957 (Bennett, Shriner and Brown, 1957). During several summers, we had recorded continuously for a week or more at a time the movements of crabs living under constant illumination of less than 1 lux. The animals were placed, individually, in actographs with a small amount of sea water, and their activities were registered on slow moving kymographs. Data of individual and group activity were read directly from those records. A clearly overt rhythm of levels of running was obvious day by day. Peaks of activity occurred twice per day; they were approximately 12.4 hours apart; they moved later, relative to solar time, an average 52 minutes per day. The peaks of running in the laboratory came several hours prior to the times of low tides on the crabs' home beach, Chappaquoit. That obvious tidal rhythm was expressed precisely for at least a week under the constant conditions. Figure 16 presents some of the cycles of August, 1955.

At least two other persons have reanalyzed the data we recorded in 1955 and 1956, and while not agreeing fully with all our conclusions, they, too, found that *Uca pugnax* functions in time with the tides while living under constant laboratory conditions. J. Enright (1965) has gone so far as to point out that our description of fiddler crab activity was the first which clearly documented persistent tidal rhythms.

Superimposed on the tidal cycle is a solar one of lower amplitude. It is characterized by greater levels of activity between 6:00 AM and noon than during the remainder of the day. The phases of the two persistent rhythms are linked to one another in some way. Webb and Brown (1965) and Webb (1971) proved that light-dark treatments which caused shifts in the

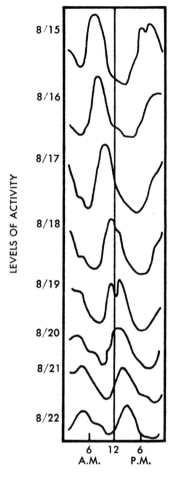

LEVELS OF ACTIVITY

6   12   6
A.M.   P.M.

TIME OF DAY

Figure 16. Overt tidal cycles of locomotor activity of *Uca pugnax* living under constant laboratory conditions, August 15 — 22, 1955. From: Bennett, M.F., Shriner, J. and Brown, R.A.: Persistent tidal cycles of spontaneous motor activity in the fiddler crab, *Uca pugnax. Biol Bull, 112*: 267-275, 1957.

phases of the circadian cycle effected comparable ones in the phases of the tidal rhythm. However, the light-dark perturbations do not entrain the tidal cycle (Webb, 1971). Earlier, we were able to shift the phases of the crabs' running cycles by exposing the animals to L/D 6/18 (Bennett and Brown, 1959). Those fiddlers also sustained shifts in the phases of their circadian and tidal color change cycles. The interrelationships between and among the rhythms of the several hands of the clock of fiddler crabs demand many more detailed analyses. Investigations of them could sharpen our understanding of timing phenomena in *Uca,* and might very well suggest additional ways of answering questions regarding environmental control and regulation of animal clocks.

An especially interesting attack by Palmer (1964) proved that the temporal behavior of fiddler crabs is tied to their orientational behavior. He recorded the degree of light preference of *Uca* which ran in actograph chambers, halves of which were darkened. The movements of the recording units indicated the times during which the animals were in the lighted ends of the chambers and those during which they were in the dark. A persistent cycle of light preference emerged. Through most of the day, the crabs reacted positively to light. Their maximal response was in the early morning. They reacted negatively to light from 6:00 PM until 1:00 AM. Once again, we see rhythmic activities to be of ecological significance. Positive reactions to light probably aid in bringing the animals out of their burrows at daytime low tides. That tendency coupled with their persistent rhythm of running at tidal frequency certainly equips them for life in the intertidal zone. My studies of the phasing of the cycles of activity illustrate how specific organismic and environmental correlates can be (Bennett, 1963). *Uca's* clock not only adapts the animals to life in the intertidal zone, but also fits them for life in a particular intertidal zone, or on a particular beach.

For my investigation, I worked with *Uca pugnax* from two different beaches, Chappaquoit on Cape Cod and Lagoon Pond on Martha's Vineyard. Those are the same areas from which Brown and his associates collected the crabs they used to compare the color change rhythms of different populations of *Uca* (Chapter 2). My animals from the two beaches were brought into the laboratory in Woods Hole on the same days. Chappaquoit, Lagoon Pond and Woods Hole all lie within a six-mile radius, but comparable tidal events come four hours earlier at Chappaquoit than at Lagoon Pond. Low tides on the former beach follow lunar zenith and nadir by one to three hours.

During the first 24 hours of recording, both groups of crabs ran maximally shortly before low tides on their own beaches. Therefore, the peaks for the Chappaquoit animals were earlier than those for the Lagoon Pond fiddlers. Phases of the 12.4-hour cycle of running had been set in time with actual tidal events. We know that phases of the tidal rhythm of color change of *Uca* are also set by tidally related Zeitgebers (Chapter 2). How is the phasing of tidal cycles accomplished? We do not have an answer to that question, but investigations by E. Naylor and his associates, now on the Isle of Man, which have been focused on the tidal rhythms of the crab *Carcinus* and which are discussed later in this chapter, have given us hints about environmental factors which set the phases of organismic cycles of tidal frequency.

A further comparison of the phasing of the tidal cycles of color change and running in *Uca*, however, proves once more how different the various indicator processes in the same species may be. The supplementary pigment dispersion appeared in time with low tides on the crabs' original beaches for at least 18 days in the laboratory (Chapter 2). The peaks of running which were seen in the laboratory an hour or so before low tides on Chappaquoit and Lagoon Pond, and were consequently four hours apart for the two groups of crabs, did not retain that relationship. Gradually, during the eight days after collection, the peaks of locomotory activity of the two groups, which were always kept in separate containers, came into synchrony, and then fell within an hour of the times of lunar zenith and nadir. That relationship was retained through the remaining four days of recording for all series that were studied. The two indicators of fiddler crabs' tidal timekeeping, color change and spontaneous activity, vary considerably in regard to their phasing mechanisms. Where do those variations lie? Are they to be found in the hands of the clocks or in their mediating pathways?

The synchronization of the phases of the overt tidal cycles of running of the two populations of crabs from the two different beaches and their coming into an obvious temporal relationship with lunar events indicate similarities in phasing phenomena in *Uca* and the American oyster. The rhythms of that mollusc were described by Brown (1954b), and are discussed in greater detail in Chapter 6. The oysters whose activities were in phase with tidal events in Long Island Sound, their original habitat, early in their sojourn in the laboratory, shifted "spontaneously" to lunar correlates characteristic of Evanston, Illinois, their place of study, after several weeks under constant conditions. What factors could have set the phases of crabs' and oysters' activity cycles in time with specific lunar events? Did that actually happen?

Enright (1965) did not agree with my interpretations of the studies of the fiddler crabs from Chappaquoit and Lagoon Pond. He contends that the coming into synchrony of the phases of the two populations was the result of the periods of the tidal rhythms of *Uca* under laboratory conditions being "slightly but consistently longer" (p. 458) than that of tidal or primary lunar cycles. If that were true, the phases of the rhythm of the Lagoon Pond animals could not have caught up in time to come into synchrony with those of the Chappaquoit animals, unless one could assume that additionally and very conveniently, the cycles of the latter crabs "waited" or stayed on precise tidal time, during the interval in which the periods of the Lagoon Pond animals were longer than 12.4 hours. That assumption is not supported by the observations for, if anything, the periods of the cycles of the Chappaquoit animals shortened a bit. Consequently, the peaks of activity came into the relationship with lunar nadir and zenith in the manner which was described. To be sure, there is some variation in the period of the running cycles under constant laboratory conditions, but a consistent increase in length is not typical.

Webb and Brown (1965) were able to relate the specific length of the period of the primary lunar or tidal cycle of activity in *Uca pugnax* to the phase of the average circadian rhythm of running during which the tidal maximal fell. As has been emphasized, the solar and lunar rhythms of *Uca pugnax* are linked.

The nature of that association is not completely explicable at present.

Some details of the running rhythms of fiddler crabs and their linkages with one another are associated with species and ecological differences. Barnwell (1963) has given us some comparative information about activity cycles. In Brazil, he recorded locomotion of *Uca maracoani* and *Uca mordax* under constant low illumination in a laboratory of the Museu Paraense "Emilio Goeldi" located in Belém. *Uca maracoani* had been collected from a tidal mud flat near Salinópolis, about 90 miles northeast of the museum, while collections of *Uca mordax* were made from ditches running along tidal streams in the vicinity of Belém.

In terms of the Brazilian crabs' environments and their dominant cycles of running, a seemingly adaptive correlation was seen. *Uca maracoani* came from the tidal mud flats, areas where the ebbing and flowing of the waters are obvious. Even in the laboratory, it ran with the tides, and showed only slight circadian fluctuations. *Uca mordax*, of the drainage ditches of tidal streams near Bélem, lived in burrows so far from the edges of the streams that they were probably covered by water only every 14 to 15 days during the semilunar high, high tides. They ran primarily at solar frequencies in the laboratory. Do they, in their natural habitat, depend upon circadian changes, rather than tidally related stimuli, to entrain and phase their rhythms of running?

Barnwell showed convincingly that some features of the cyclic patterns of locomotion of Woods Hole fiddler crabs are associated with day-night changes, specifically changes in the photoperiods to which the animals were exposed (Barnwell, 1966). He recorded the activity of groups of *Uca pugnax*, *Uca pugilator* and *Uca minax* under constant (L/L) low illumination and while exposed to natural day-night (L/D) changes of July and August. In L/D, all species had precise tidal rhythms for as long as 46 days. Superimposed on the 12.4-hour cycles were ones of circadian frequency. The normal diurnal changes apparently helped the animals maintain the constancy of their tidal cycles. Supporting that theory are these observations. When the crabs were in constant light (L/L), their running rhythms changed by the end of a 14-day period. The period of the cycle of *Uca minax* exceeded 12.4 hours. The cycles of *Uca pugnax* and *Uca pugilator* became strongly circadian. Explanations of those observations are also lacking.

But the observations, themselves, reemphasize how complex the relationships between solar and tidal timekeeping are. The current investigations of H. M. Webb (1971) on the effects of artificial 24-hour cycles on the tidal rhythms of activity in *Uca pugnax* promise pointed information about the linking of the crab's cycles of the two frequencies. Her work emphasizes, too, that phase-setting and entrainment may not be so closely associated physiologically as has been assumed. Twenty-four hour L/D treatments shifted phases of the tidal cycles of *Uca*, but, in no case did the solar-day cycles entrain the tidal cycle. The interrelationships of the organismic rhythms at the levels of the clockworks, the mediating pathways and the indicator processes must be analyzed fully before we can claim complete understanding of the biochronometry of fiddler crabs.

Very little is known about the mediating pathways of the cycles of running activity or of oxygen-consumption of these crabs. Only suggestions that neurosecretions of the eyestalks influence the expression of the metabolic cycles are available (Brown, Bennett and Webb, 1954). The suggestions are reinforced by observations and experiments on two other crabs, *Gecarcinus lateralis* and *Carcinus maenas*. Dorothy Bliss of The American Museum of Natural History has questioned the land crab, *Gecarcinus*, about possible relationships among stages of molting, concentrations of eyestalk neurosecretions and locomotor activity (Bliss, 1962). In its natural tropical environment, the crab is nocturnal. After being entrained to an L/D 12/12 schedule and then being placed in constant darkness, *Gecarcinus* continued to be most active at circadian frequencies. Changes in that rhythm were seen in animals from which the eyestalks had been removed and in crabs which had entered the premolt stage, a time during which the concentrations of one secretion of the eyestalk complex, the molt-inhibiting hormone (MIH), are normally low in the body fluids. The levels of activity of those two groups of crabs were lower than the levels of normal *Gecarcinus* not approaching molt. Also, the frequencies of the circadian rhythms of eyestalkless and premolt animals were much more variable than were those of nonmolting crabs.

In these two respects, levels of activity and variations in period length, the cycles of normal premolt and eyestalkless crabs were very similar to each other and commonly different from rhythms of nonmolting crabs. Incidentally, these findings convince us that in addition to knowing the ecology and the immediate past history of the organisms with which we work, students of biochronometry should know the developmental and physiological states of the experimental subjects. It is apparent in *Gecarcinus* that decreases in the titers of MIH, whether induced by the removal of the eyestalks or by the withholding of that neurosecretion from the circulation during the premolt stage, influence the crabs' circadian rhythm of locomotion. Is that neurosecretion part of the mediating pathways of the cycle? Does it exert its influence on the cellular clockworks?

Neurosecretions of the eyestalk complex are mediators of the activity cycles of the green crab, *Carcinus maenas*. E. Naylor and his associates, working at the University of Swansea, at the Stazione Zoologica in Naples and on the Isle of Man, have published a series of reports on rhythms in this crustacean. Specimens from intertidal zones show tidal, circadian and semilunar cycles of running when maintained in the laboratory under constant conditions (Naylor, 1958). The peaks of the tidal cycles are in synchrony with the times of high tides in the animals' home habitats. Those are the periods during which *Carcinus* moves about in the sea. It hides beneath rocks at low tide. The frequency of the tidally related cycle is very close to 12.4 hours. The circadian rhythm of *Carcinus* is characterized by maximal running during the night, while the fortnightly cycle reaches its highs on solar-days within which high tides fall during darkness. As have many of us, Naylor concluded that circadian and tidal cycles are linked, and their simultaneous possession produces organisms' semilunar rhythms.

Palmer (1967) described very similar persistent rhythms for

still another intertidal crab, *Sesarma reticulatum.* It is often found
on Cape Cod burrowing in salt marshes near where *Uca pugnax*
lives. *Sesarma,* like *Carcinus,* was more active during the night
than during the day, and at times of high tides rather than lows,
when it lived under constant conditions. Thus, the phases of the
persistent cycles of these two species are the "reverses" of the
rhythms of the fiddler crabs.

The investigations by Naylor's group (Naylor, Atkinson and
Williams, 1971 and Naylor and Smith, 1973) have focused on
points which have also been approached via studies of the fiddler
crabs' persistent cycles of running. The agreement between the
crabs' patterns of activity in the laboratory and those in their
native habitats have been cited. As emphasized, the maxima of
the persistent tidal cycles of the intertidal *Carcinus* were synchronized with the high tides on the coasts from which the
animals were collected. *Carcinus maenas,* captured in the dock
areas of Swansea where the tides are negligible, also ran at 12.4-
hour frequencies when living in constant light. However, when
the dock crabs were exposed to a day-night lighting regime, they
were most active at circadian frequencies with their peaks of
locomotion coming during the dark periods.

A change from an expression of a primarily tidal pattern
under constant illumination to one of a circadian nature under
L/D conditions was also seen in crabs collected from an open
coast and then kept for nearly four weeks in aquaria. Do all
Carcinus have the capacity to function at tidal frequencies, but
is that behavior not overt when the animals remain in nontidal
environments?

Perhaps all *Carcinus maenas* have tidal clockworks, but
Carcinus mediterraneus from the Bay of Naples seem not to
possess them. There, there is virtually no tidal flow. Crabs, collected from the Bay and maintained in the Stazione Zoologica,
were most active at 24-hour frequencies even under constant
light. Their maximal running also occurred during actual nighttime. The tidal periodicities typical of the British crabs, whether
from open coasts or from the docks, were not even suggested in
the performances of the Neopolitan animals. *Carcinus* has also

taught us that some characteristics of persistent rhythms are species-specific.

Species differences were obvious in the results of studies of temperature change on the cycles of the Italian and British *Carcinus*. The locomotory rhythms of all groups were the same at 10, 15, 20 and 25° C, and the movements of all were suspended at 4° C. However, after the animals had been chilled and returned to constant conditions which included higher temperatures, the variations among the behaviors of the shore and dock *Carcinus maenas* and *Carcinus mediterraneus* were recorded. No matter at which real time chilling was initiated, as long as it was continued for more than six hours, a rephasing of the cycles of the shore *Carcinus* was the result. A burst of running occurred when the ambient temperature was raised. High levels of locomotion were then repeated at 12.4- and 24-hour frequencies. The phases of animals' typical persistent cycles had merely been shifted by the temperature perturbations.

The rhythms of the dock *Carcinus* after chilling were the same as those of the shore crabs after identical treatment. But, for the dock or nontidal forms, the results did not indicate mere rephasing, for before chilling, the crabs had run at only solar frequency, and after having been in the cold, they ran at both 24- and 12.4-hour frequencies. In the dock *Carcinus*, the temperature perturbations seemed to induce the expression of the tidal rhythm and to shift phases of the circadian cycle.

Locomotory patterns in which tidal peaks were emphasized were also evoked in laboratory-reared *Carcinus maenas* by chilling at 4° C. The young crabs moved about at only a 24-hour frequency prior to their six-hour exposure to the cold. When their temperature was raised after chilling, they, as did the mature shore and dock crabs, ran in time with the day and the tides. In the animals' natural habitats are the phases of the 12.4-hour cycles set by temperature changes? At least during the winter, the British crabs are probably exposed to temperatures which approach 4° C. But it would seem unlikely that animals in the Bay of Naples would ever be exposed to such lows. What was their response to chilling in the laboratory? None. Their persistent rhythms of running were the same before and after periods of low temperature. The cycles continued to be solar ones with no hints of phases which correlated with tidal phenomena. Does *Carcinus mediterraneus* lack the genetic capacity to run in time with the tides? Does it, perhaps, demand different entraining or synchronizing factors, *e.g.,* mechanical stimuli of moving water, to cause a latent tidal rhythm to become overt?

Perturbations associated with the ebbing and flowing of the tides certainly affect tidal timekeeping in *Carcinus maenas.* Naylor collected crabs from the docks of Swansea, caged the nontidal crabs and took them to the open coast where they were set between or below the tide marks. They lived there for eleven days after which their locomotion was recorded under constant laboratory conditions. Tidal and circadian rhythms were obvious. Cycles of the same general, but less precise, natures were recorded after dock crabs lived in the tidal zone for only two days. Which of the physical changes of tidal flux — movements of the water, changes in hydrostatic pressure, changes in temperature — were Zeitgebers for the animals moved to the coast? In the laboratory, Naylor has succeeded in effecting the expression of the tidal cycles of dock *Carcinus* by chilling for six hours, as described earlier; by exposing the animals in moist air to five days of temperature change between 13 and 24° C (an 11 degree differential) at tidal frequencies; by five days of 6.2 hours in air at 24° C followed by 6.2 hours in water at 13° C (again, an 11 degree difference); and by five days of 6.2 hours at 17° C in air alternating with 6.2 hours at 13° C in water (only a 4 degree temperature difference). The same regimens shifted the phases of the overt tidal rhythm of *Carcinus* brought in from the coast.

Neither group of crabs, dock nor shore, responded to exposures to 4° C of less than six hours; to tidally cyclic temperature changes of only 4 degrees difference in moist air; to treatments consisting of 6.2 hours in water followed by 6.2 hours in air at the same temperatures. On the basis of these findings, one tends to postulate that also in its natural habitat, *Carcinus* uses a combination or group of variables to phase its rhythms of activity with those of its physical surroundings. Additionally, one must assume that the crabs which did respond to the experimental manipulations possessed the genetic capacity to be entrained by treatments of tidal frequency. What would result if *Carcinus mediterraneus* were set between tide lines or were subjected in the laboratory to the combinations of factors which effected or affected tidal rhythms in the British crabs? It is to be recalled that chilling the Italian crabs for six hours did not evoke running at tidal frequencies under constant conditions.

Other crabs found to be unresponsive to the changes which influenced the tidal rhythms of animals from Swansea were *Carcinus maenas* from which the eyestalks had been removed. Mentioned earlier in this discussion was Naylor's conclusion that neurosecretions of the eyestalk complex are involved in the mediation of the persistent activity rhythms of green crabs. His evidence has come from several different experimental attacks. Firstly, as Bliss did with *Gecarcinus* and as we did with *Uca,* the British investigators compared the patterns of locomotion of de-stalked and sham-operated *Carcinus maenas* under constant laboratory conditions. The sham-operated crabs showed the precise 12.4-hour and 24-hour cycles which have been described; the animals deprived of their eyestalks were arrhythmic. Furthermore, the levels of activity of the experimental crabs were significantly greater than those of the animals with eyestalks. Does a chemical factor of the eyestalk tend to inhibit locomotion? Naylor made that suggestion, attacked the proposition experimentally and secured data which supported it.

He injected eyestalkless *Carcinus* with extracts of eyestalks removed from donors during their periods of rest or minimal activity. The eyestalkless recipients became much less active. Is it possible that effective Zeitgebers influence the green crab's tidal rhythm via the eyestalk complex, itself? Yes. Keeping the crabs at 4° C for at least six hours serves to establish or to rephase 12.4-hour cycles in *Carcinus maenas,* and chilling only the eyestalks while the remainder of the crab's body is kept at a higher temperature also results in the establishment of typical tidal rhythms of running. Finally, Naylor has found that injections of eyestalk extract into normal or eyestalkless crabs succeed in entraining the 12.4-hour cycle. Neurosecretions are definitely implicated in the mediating pathway of the persistent cycles of locomotion in *Carcinus*.

Whether secretions of the eyestalks or of other neural tissues mediate additional cycles in *Carcinus* is another question. Arudpragasam and Naylor (1964) discovered that the levels of gill ventilation and of oxygen uptake also vary at tidal frequencies in *Carcinus maenas* which had lived on open coasts exposed to tidal flow. Only a hint of a superimposed circadian rhythm was discerned. Comparable phases, *i.e.*, maxima and minima, of the two physiological tidal cycles paralleled one another, but were not in synchrony with the highs and lows of the crabs' running rhythm. The tidal peaks of respiratory activities occurred later than those of the locomotory cycle. That temporal relationship resembles closely one found for the rhythms of oxygen-consumption and motor activity in *Uca pugnax*. Additional comparative studies of the several indicator processes of many individual species are necessary to clarify the complexities which exist among the hands and mediating pathways of any one set of cellular clockworks.

Comparative investigations stressing phylogenetic and ecological variables would also be of importance to students of biochronometry. The class, Crustacea, presents an excellent collection of creatures for such studies. This discussion has focused upon the clocks of but a few of the true crabs. There are prawns whose retinal pigments migrate with circadian frequency; there exist littoral amphipods that orient by time-compensated compass reactions; there are beach dwelling isopods which swim and burrow with tidal frequency (Sollberger, 1965). Experimental investigations of the tidal cycle of the beach fleas are two of the very few which have yielded evidence that persistent cycles can be influenced by chemical manipulation (Enright, 1971a and 1971b). The free-running period of the isopods' activity rhythms lengthened after treatment with deuterium oxide and with ethyl alcohol. The mechanisms of these effects have not been explained.

Further, cyclically recurring changes in fresh water and cave dwelling crayfish have been described. The ancestors of these animals left the sea, and, in the case of the cavernicolous species, left the light of day, millions of years ago. Yet, the blind cave crayfish, *Orconectes pellucidus* ticks off the time of day (Brown, 1961 and Jegla and Poulson, 1968) and the time of year (Jegla and Poulson, 1970) and its sighted, surface dwelling relatives, *Orconectes clypeatus* (Fingerman and Lago, 1957) and *Cambarus virilis*, (Guyselman, 1957) also function at solar frequencies. The locomotory activities of *Orconectes pellucidus*, collected from Mammouth Cave, Kentucky, and observed under constant conditions in 1935; *Orconectes clypeatus*, found and studied in Louisiana in the winter of 1956; and *Cambarus virilis*, collected and maintained in the laboratory in Minnesota in the summer of 1956, varied through the solar-day in amazingly similar patterns. All species moved about at minimal levels between late morning and midday, and all were maximally active around 6:00 PM (Fig. 17).

Maxima during subjective nighttime were also obvious in the circadian cycles of bioelectrical changes in the nervous system of crayfish (Aréchiga, Fuentes and Barrera, 1973). Rhythmic activity in the sustaining fibers of the optic nerves were recorded in L/L, D/D and L/D for as long as eight weeks at a time, and the spontaneous changes as well as responses of the fibers to light pulses were greatest during the night. Similar phasing was seen in the cycles of electrical changes in receptor and interneurons, and in the circadian rhythms of retinal pigment migration and locomotion in the same species of crayfish, *Procambarus clarkii* and *Procambarus bouvieri*, forms which are nocturnally active in their natural environments.

All these cycles of the crayfish appear to be mediated chemically in a manner which parallels that postulated for *Gecarcinus*, *Carcinus* and *Uca*. Aréchiga has been able to induce daytime phases of the crayfish rhythms by injecting extracts of whole eyestalks or of only sinus glands, and has found that the circadian variations disappear in animals from which the eyestalks or the brain have been removed. The results of Aréchiga's experiments suggest that the brain evokes the release of sinus gland secretions through some neurosecretory mechanism.

The electrophysiological techniques used by Aréchiga in his analyses of crayfish rhythms are some of the most difficult, but

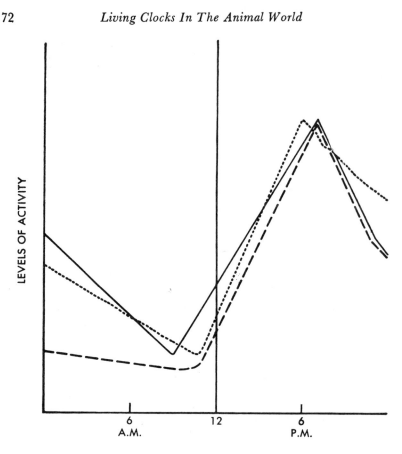

LEVELS OF ACTIVITY

6
A.M.

12

6
P.M.

TIME OF DAY

Figure 17. Comparison of the solar cycles of locomotor activity of *Orconectes clypeatus* (dotted line), *Orconectes pellucidus* (solid line) and *Cambarus virilis* (dashed line). From: Fingerman, M. and Lago, A.D.: Endogenous twenty-four hour rhythms of locomotor activity and oxygen consumption in the crawfish *Orconectes clypeatus. Am Midl Nat,* 58: 383-393, 1957; Brown, F.A., Jr.: Diurnal rhythm in cave crayfish. *Nature, 191:* 929-930, 1961; Guyselman, J.B.: Solar and lunar rhythms of locomotor activity in the crayfish *Cambarius virilis. Physiol Zool,* 30: 70-86, 1957. Copyright 1957 by the University of Chicago Press.

most precise, approaches to our unsolved problems of bio-chronometry. Their application to investigations of the clocks of the true crabs as well as to studies of timing in all animals would seem especially promising.

Chapter 4

HONEY BEES

FOUR YEARS BEFORE the outbreak of World War I, the German edition of August Forel's book on the senses of insects was published in Munich. In it, Forel, a Swiss physician and naturalist, claimed for honey bees a "Zeitgedächtnis" or time-memory. Twenty years after the cessation of World War II, the manuscript of Karl von Frisch's book on the dance language and orientation of bees was finished in Munich. In it, von Frisch, the world's authority on honey bees, made the same claim, and used the identical word, "Zeitgedächtnis" and a related one, "Zeitsinn" or time-sense. Forel's claim was based on casual observations made from the terrace of his summer home in the Swiss Alps. von

Frisch's was based on precisely planned studies made in his laboratory in Munich and in the meadows and mountains near his summer home in the Austrian Alps.

During the half-century between the publications of those two books, von Frisch, his associates and his students, who have contributed so great a part of the fascinating story of the lives of honey bees, had proved that these insects have an "Innere Uhr" (internal clock), and that its indicators, time-memory and rhythmically repeated activities, are integral parts of several complex behavioral patterns of the bees.

One interested in seeing for himself the functioning of honey bees in time with the cycles of their natural environments need do only as Forel and von Frisch did so well — watch bees. During the daytime of warmer parts of the year, one can observe workers collecting pollen, nectar and water and delivering them to the hive. In the spring and summer, when particular trees, shrubs and flowers are in bloom, an observer can watch the bees come and go between the blossoms and the home colony on fair days.

If one extends his observations past sunset, he can prove immediately that honey bees are day active, for few or none visit sources of food after the light intensity falls with approaching night. The careful observer can also note the frequency of visits of bees to the blossoms under surveillance. He usually finds that they increase after the first few visits of the morning. By watching bees which have been markèd for individual identification, one can make such observations quantitative. Shortly after marked bees fly back to their hive from the flowers, unmarked or newly recruited workers arrive in greater and greater numbers. Those, too, can be marked for identification, and the rates and times of their trips to the sources of pollen and nectar can then be scored.

Finding and collecting nectar and recruiting additional workers by communicating to them the distance and direction of its source from the hive are activities now well-known and understood — thanks again to elegant investigations by von Frisch and his associates. Some of our finest examples of behavioral analyses of receptor phenomena, animal communication, and orientation, some of which is time-compensated and, therefore, demands an organismic timepiece, are to be found in the literature of the biology of the honey bee (von Frisch, 1967).

However, it is only fair to say that the observations by Forel and his family provided the incentive for the characterization and explanation of timing in honey bees. As narrated in his book on insect senses (Forel, 1910), early summer breakfasts were enjoyed on the terrace of the Forel home in the mountains. The family noted a few worker bees from a hive located about 120 steps from the house sampling marmalade and other sweets from the breakfast table. Before long, more and more bees came every morning. As the days progressed, the bees often appeared on the terrace even before breakfast was served, as if awaiting or expecting a reward.

Finally, the Forels moved breakfast into the house, for it was impossible to eat accompanied by the great number of bees which flew in every morning, and only in the morning. However, bees continued to arrive on the terrace at breakfast time for several days longer even though there was no food there to attract them. It was described as "amusing" to watch them crawl about on the table to seek "obstinately" for sweets. Since that searching was observed only once a day around the solar time at which food had been present earlier, Forel postulated that the bees possessed a memory for time.

The word, "Zeitsinn" or time-sense, was first used by a German zoologist, von Buttel-Reepen (1915), in his description of the temporally precise visits of honey bees to fields of buckwheat. He reported that bees came to those brightly colored, fragrant blossoms only between approximately 10:00 AM and 11:00 AM. During that interval, buckwheat secretes its nectar. During midday and afternoon, though the weather remained fair and the fields bright, honey bees did not appear. But in the mornings, the workers were seen collecting industriously and always between 10:00 and 11:00 o'clock.

Because of the several possible explanations for those actions, few hints of the basic nature of the bee clock were forthcoming from the reports of Forel and von Buttel-Reepen. In both cases, the worker bees could have used environmental cues as Zeitgebers. The bees observed in the buckwheat fields could have responded directly to the nectar, and therefore, would not have flown to the flowers when that sweet material was lacking.

Evidence for a time-memory or time-sense based on an internal clock was published in 1929 by Beling, a student of von Frisch. Her work was done in the laboratory under constant conditions of light and temperature. She used an observation colony of *Apis mellifera,* members of which were marked as they collected sugar water. Collection was from a glass container located 10 meters from the hive. Sugar water was available only between 3:00 PM and 5:00 PM, but all other aspects of the feeding table and collecting dish were kept the same through the entire solar-day. Workers collected between 3:00 PM and 5:00 PM (the training time) for four or five consecutive days. During the following days (the observation or test days), no sugar water was present on the table. The investigators sat next to the empty collecting dish where they recorded the times at which the bees whose identity could be noted searched the table. Figure 18 is the type

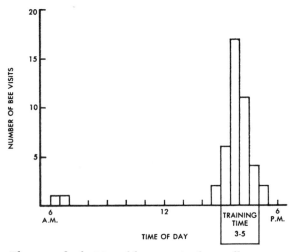

Figure 18.   The record of visits of bees, trained to collect sugar water between 3:00 and 5:00 PM, on a test day when the collecting table was empty. From: Beling, I.: Über das Zeitgedächtnis der Bienen. *Z Vergl Physiol,* 9: 259-338, 1929, Berlin-Göttingen-Heidelberg: Springer.

of graph used to illustrate the temporal patterns of visits to the collecting areas on observation days. It shows that most visits to the collecting table occur shortly before and during the training time. Only a few bees are there at other times. That pattern is repeated, usually for six to eight days after the training period, and the interval between points of maximal frequencies of visits is very close to 24 hours. Beling and others have also shown that the times of the most intense activity at the collecting table can be set at various hours of the actual solar-day, so that event is not locked to a particular one of the physical environment.

The phasing depends upon the training time. If bees are trained to collect for several days between 3:00 PM and 5:00 PM, on observation days, they come to the empty collecting dish just before and during that interval. On the other hand, if the training time is between 10:00 AM and noon, then the greatest number of visits to the collecting site will occur between roughly 9:30 AM and 12:30 PM. Workers can also be trained to collect vigorously during several 1- to 2-hour intervals of the day if at least two hours of noncollecting time are interposed between the periods of collecting. In fact, the bees of one hive can learn to seek nectar at one place during one training period and at another site during a second period. These patterns, too, are maintained or remembered for six to eight days after the cessation of training. The internal clock responsible for the time-memory is circadian. Training at intervals of 19 or 48 hours miscarries. Workers trained at 19-hour intervals visit the collection table almost constantly on observation days. Those which learn to collect sugar water every 48 hours, seek it at 24-hour intervals during periods of observation.

Beling, as well as Wahl (1932), who repeated and extended the earlier studies of the bees' time-memory, sought to establish whether or not environmental factors whose levels varied with solar frequency signaled phases of circadian time which might provide time cues for the bees. Their observation hives were generally kept under constant low illumination, constant temperature and constant humidity. Wahl repeated his work in a salt mine in the Bavarian Alps, and Beling counteracted changes in the levels of ionization of the atmosphere of the laboratory with radioactive materials. In all cases, the bees which had been trained by being allowed to collect sugar water for several hours every day, returned to seek the sweet solution at 24-hour intervals. Those investigators were convinced that *Apis* possesses an internal clock which is inherited, is not learned in the hive, and is independent of external environmental factors for its determination of frequency.

Questions of the effects of temperature and drugs on circadian rhythms were also put to the honey bee. The period of the collecting cycle is temperature-independent between 5° and 38° C. Bees do not live at temperatures which exceed 38° C, but those which are trained, and are then cooled at 4° to 5° C for several hours, arrive at the collecting dish late by the same number of hours as those of cooling (Fig. 19). Although earlier investigators recorded some aberrations in the collecting cycle after trained bees had been treated with quinine or thyroxine, careful repetition of those experiments (Renner, 1961) yielded no evidence of any effect of the drugs on the frequency of visits to the collecting table.

In addition, worker bees, trained and then treated with ether, sought to collect on time. There are no records of attempts to influence the bee's clock or its hands by treating the animal with extracts of parts of the central nervous system of another individual of the same species. Is it not possible that intrinsic neurosecretions might have effects on the form, amplitude or period of rhythms of bees even though thyroxine does not? There is very little evidence of any influence of vertebrate hormones on invertebrate animals. Neuroendocrinologists might well find some answers to questions of integration and regulation in the Hymenoptera by analyzing the mediating pathways of the internal clock of the honey bee.

Recently, Medugorac and Lindauer (1967) proved that CO_2 narcosis affects the time-memory of bees, and thereby supported earlier claims of Kalmus (1934). They trained bees to collect in the usual manner for five days, and after the last training period, the animals were treated for four hours with 20 per cent CO_2. During narcosis, the animals were quiet although weak motor

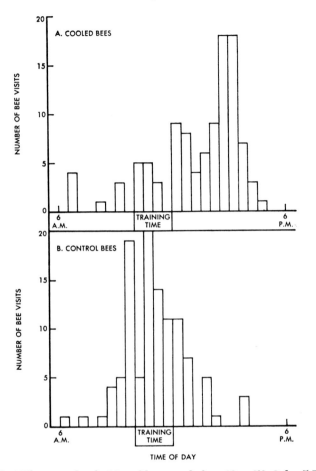

Figure 19. The records of visits of bees cooled at $4° - 5°$ C for 5.75 hours before the test day (A.) and of their controls (B.). All the bees had been trained to collect between 10:00 AM and noon. From: Renner, M.: Neue Versuche über den Zeitsinn der Honigbiene. Z *Vergl Physiol, 40*: 85-118, 1957, Berlin-Göttingen-Heidelberg: Springer.

responses could always be evoked, but within ten minutes of the end of treatment, they began to fly. The pattern of visits to the collecting table on the observation days was a strange one. There were two intervals of maximal seeking, one during the two-hour training period and a second about three hours later. The same bees participated in the activity of both periods. Whether the hive was kept under constant low light intensity or was exposed to light-dark changes, the two-peaked pattern was seen.

The investigators claim those results as proof that the inner clock of the bees consists of at least two components, only one of which is influenced by CO_2 narcosis. The part not affected evokes the collecting 24 hours after the training time, while the second part which does respond to the treatment with CO_2 is responsible for the period of collecting which is late by roughly the same amount of time as that of narcosis. Evidence for two components was also adduced from the results of investigations by Beier (1968) on light-dark perturbations and phase-shifting of the collecting cycle of honey bees. What is the physiological difference between those two components of the bees' internal timing mechanisms? Perhaps one is more sensitive to environmental factors or Zeitgebers than is the second. Studies of the role of social Zeitgebers and the two-component cellular timepiece are discussed later in this chapter.

Such precise studies of the bee's time-memory and its mechanism were made possible by the development by Max Renner of the Institute of Zoology, University of Munich (Renner, 1955a), of techniques with which a colony of honey bees can be maintained in a normal, healthy state for long periods of time under constant conditions of light, temperature and humidity. A colony in an observation hive can be kept in a room, called simply a bee laboratory (Fig. 20), 7 m long by 3 m wide by 3 m in height, lighted diffusely at an intensity of 1000 lux. The temperature must be maintained close to 28° C, and the relative humidity should vary only narrowly between 60 and 80 per cent. The walls, floors and ceilings of those laboratories are painted white. Large figures, *e.g.*, X's and triangles, used by the worker bees in their spatial orientation are painted on the walls. Sources of water, pollen and sugar water are also necessary. Under these conditions, most colonies will thrive, with a fertilized queen laying eggs, the workers caring for the brood and collecting vigorously when sugar water is available.

Using the methods worked out by Beling, one can train bees to collect from particular places in the laboratory during specific

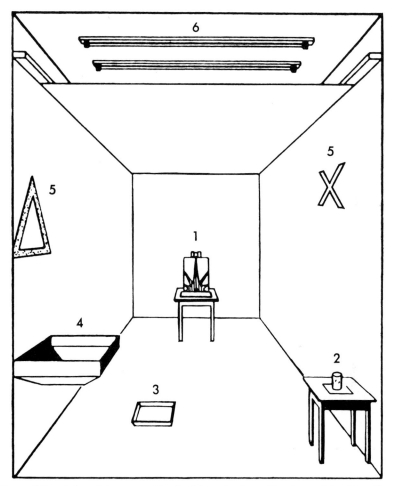

Figure 20. A bee laboratory. *1*. the hive; *2*. the collecting table; *3*. pollen; *4*. drinking water; *5*. orientation markers on white walls; *6*. lights. Two of these were built to be shipped and reassembled for the New York-Paris (see text) experiment.

periods of the day. Environmental factors such as photoperiod or temperature can be modified or changed within the bee laboratories, and the laboratories, themselves, can be constructed in such a manner that they can be disassembled, packed, transported and reassembled in new localities. Two identical bee rooms, built in Munich and shipped, one to Paris and one to

New York City, were used by Renner and his assistant when in 1955, they were able to carry out an experiment first conceived by von Frisch. As was discussed in Chapter 2, that study asked the same question we asked of fiddler crabs collected from Cape Cod and flown to California. Can the organismic timepiece maintain its frequency when the living system is transported rapidly through several time zones?

In 1937, von Frisch pointed out that most of the results of the investigations of the clock of the bee supported the notion that it was endogenous, for the animal's time-sense did not depend upon recurring changes in factors of the external environment. It was probably wholly metabolic. However, he emphasized that that idea had not been proved, and could not be proved so long as the animals were studied in one time zone. There, geophysical factors whose levels vary with the rotation of the earth, *e.g.*, gravitational levels, geomagnetic fluxes, could not be screened out completely, and might provide time cues for the bees which learned to collect sugar water at 24-hour frequencies. Even though one could not screen out the changes or even compensate for them, he could vary the times at which the animals were exposed to specific phases of solar and lunar cycles by transporting the bees, very rapidly, to different longitudes on the face of the earth.

If bees were trained to collect between 10:00 AM and noon in Paris, and after their last training period were flown to New York City, where most rhythmic geophysical events occur five hours later than they do in Paris, would they seek sugar water on Paris or New York time? If their maximal activity at the collecting station occurred between 10:00 AM and 12:00 noon, New York time, or 29 hours after their last collection in Paris, evidence of an external timer could be claimed. If the bees continued to fly to the training table between 10:00 AM and noon, Paris time, or 24 hours after their last training period, greater support for a completely endogenous internal clock would be gained.

The outcome of that classic experiment is well-known (Renner, 1955b) and parallels that of our translocation study using rhythms of color change in *Uca* as indicators of living clocks. The 40 worker bees, trained to collect between 8:15 PM and 10:15 PM, French Summer Time, were flown overnight to New York City. With the cooperation of the staff of the airline, the United States Department of Agriculture, the Customs and Immigration Services, the New York Police Department and members of the staff of the American Museum of Natural History, Renner carried his bees into the bee room which had been set up in the museum, less than 20 hours after they had been shut in their traveling hive in Paris. That bee room, its light intensity, humidity and temperature were exactly the same as those in Paris. The bees answered von Frisch's original question neatly and promptly. They flew to the collecting table with greatest frequency between 3:15 PM and 5:15 PM, Eastern Daylight Time, or 24 hours after their last training time in Paris. Their clocks had remained on French Summer Time.

The experiment was reversed by retraining workers on New York time, and then flying them back to Paris. Those results were comparable to the ones recorded in New York. The bees attempted to collect sugar water 24 hours after their last sampling of it in New York City. In Paris, now, they were still living on Eastern Daylight Time, and did so during three days of observation. Cycles of unidentified external factors had not disrupted, reset or modified the endogenous timing mechanism of *Apis*.

However, as the work of Beling, which has been described, showed, the phases of the collecting cycle of honey bees, or the hands of their clock, can be set in time with various points in the solar cycle. As will be considered again later, the position of the sun in terms of its azimuth (direction toward the sun on the horizontal plane, and measured in degrees clockwise, starting from North), not its altitude, is an effective Zeitgeber. It was pointed out earlier that the roles of the sun's position and the bee's internal clock in the animal's orientation to sources of food and the hive, as well as in their communication of the location of nectar to other workers of the colony, were elucidated by von Frisch and his colleagues. The bees use the position of the sun or the position of its pattern of polarization as a compass point, and maintain their flight line between the collecting areas (food) and the hive at particular angles with that position (Fig. 21).

However, since their compass point moves as the earth rotates, the angle between their goals and that compass point must change at approximately the same rate (15 degrees/hour) as the apparent movement of the sun's azimuth, if they are to orient successfully. They do orient successfully for their internal

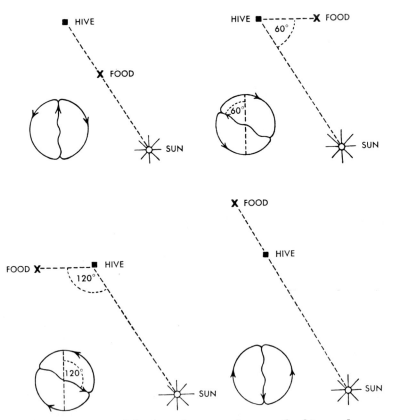

Figure 21. Diagrams of the lines between the sun, the hive and sources of food at four different times of day, and the patterns of "waggle" dances performed by workers inside the hive at those times of day. Adapted from Karl von Frisch: *Bees: Their Vision, Chemical Senses, and Language*, rev. ed. Copyright 1950, ©Copyright 1971, by Cornell University. Used by permission of Cornell University Press.

clocks allow them to compensate for the movement of the sun at roughly the proper rate. *Apis*, as some birds and beach fleas are able to do, moves in its environment in a time-compensated manner or shows what we have come to call clock-compass orientation.

Compensation for the change in the sun's azimuth is also apparent as successful foragers recruit new collectors by indicating the location of nectar and pollen. The communication of that information depends primarily on the famous "waggle" dances of the bees which are performed on the vertically arranged combs inside a hive. The "waggle" dances are described fully and dramatically by von Frisch, himself, in his book of 1967, *The Dance Language and Orientation of Bees*.

In 1970, Gould, Henerey and MacLeod confirmed the original hypothesis that a forager's "waggle" dance conveys directional information. Their work was prompted by erroneous, but loud, criticisms of von Frisch's ideas. During a "waggle" dance, the angle between the line of the forward, "waggling" part of the dance and "up" or directly away from the force of gravity indicates to the recruits the angle which exists between the sun's azimuth and the line of flight to the rich supply of nectar (Fig. 21). The dancer is translating the position of the sun into the position of gravitational force. The dance angle, too, must and does change during the day as the position of the sun, relative to the hive and sources of food, changes. von Frisch noted only slight errors in the bees' translation, and we know now that those are influenced by geomagnetism (Lindauer and Martin, 1968).

Sometimes, a bee will dance all night, and as she does, she "calculates" the course of the sun on the other side of the earth and indicates that, as usual, in the "waggle" run of the dance (Lindauer, 1971). Such dancing occurs when scout bees communicate the location of a possible new home to a swarm which has left its original colony. During those extended performances, one can watch the angle between the dance line and gravity change gradually, again at approximately 15 degrees per hour. This "marathon dance" illustrates obviously and clearly the remarkable precision of the timepiece of the bee. The adaptive nature of the capacity to function at a circadian frequency is probably nowhere more pointedly emphasized than in the lives of honey bees, animals whose existence depends upon locating food sources, finding the home colony after long flights, and recruiting others to help collect during the warm parts of the year. Their food sources, themselves, depend upon the sun; their orientation to those sources and their communication of the location of them depend upon the position of that body; the animals' lives proceed in time with the solar cycle.

The adaptive relationship between the bee's internal clock and the azimuth of the sun, as a Zeitgeber, also became apparent in the results of a second translocation study conducted by Renner and Daumer in the summer of 1957 (Renner, 1959). That series of observations was made in the field, near St. James on Long Island, and near the University of California at Davis. The two fields, selected for their great similarity in terms of landmarks, are roughly 45 degrees longitude, or three hours, apart. The investigation questioned the effects of external factors on the searching performance of the bees which had been shown to depend primarily on an endogenous clock. On Long Island, workers were trained to collect during one period, between roughly 1:00 PM and 2:30 PM, Eastern Daylight Time, at a station which was about 150 yards northwest of the hive. The collecting stations of that investigation were automatic recording ones which registered the times at which bees entered and left the chambers in which sugar water was located.

After the plane flight to California, the hive was set in the center of the open field, and eight automatically recording units, exactly like those used for training in New York, were set 150 yards from the colony and 45 degrees away from each other. Therefore, eight points were equipped with instruments which would register the visits of bees searching in any of eight compass directions. The foragers were tested for three consecutive days, and on each of those days, two peaks of searching activity were recorded. Figure 22 illustrates a two-peaked pattern which had not been observed in studies made prior to Renner's, and that the times of the peaks changed slightly from day to day.

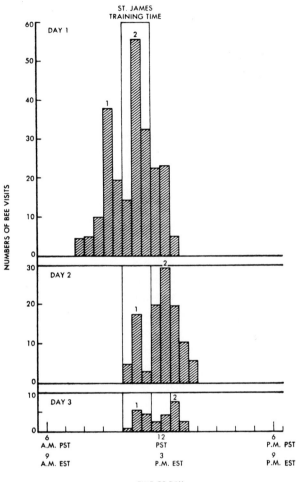

Figure 22. The records of visits of bees, trained to collect between 1:00 and 2:30 PM, E.S.T., on Long Island, on test days 1, 2 and 3 in California. 1 and 2 indicate peaks as identified by Renner (see text). From: Renner, M.: Über ein weiteres Versetzungsexperiment zur Analyze des Zeitsinnes und der Sonnenorientierung der Honigbiene. *Z Vergl Physiol, 42*: 449-483, 1959, Berlin-Göttingen-Heidelberg: Springer.

Renner believes that the first peak (1 in Fig. 22) reflected the operation of only the cellular clockworks. Its displacement in time was only minor, and that was considered proof that the bees continued to collect in California on New York time. However, the changes in the timing of the second peak (2 in Fig. 22), which by the third day of observation had shifted by almost three hours, were thought to be of major significance, and were attributed to the actions of external factors. According to Renner's calculations, the only external factor whose temporal movement could explain these results is the azimuth of the sun, the factor which von Frisch had earlier identified as the compass point of bee orientation.

When bees are trained and are later tested in the *same* field, the maximum, the most obvious phase of the bees' cycle of searching at the collecting sites, occurs during one circumscribed period of the solar-day, the original time of training. But after the rapid transport, the period of the bees' maximal searching, as set on Long Island, did not occur in time with the same phase of the solar cycle in California as it had in New York, and the searching stimulated by the internal circadian clock occurred during the first peak, which shifted very little in three days, while the activity evoked by the position of the sun contributed to peak two which moved toward a later time by almost three hours in three days.

Comparable patterns of desynchronization of searching activity were evoked experimentally by exposing bees, working in a bee laboratory, to light-dark changes which were made 3.5 hours later on observation days than they had been made on training days (Beier, 1968). Beier believes that his results also support the idea of a two-component honey bee clock. One of the components is more sensitive to exogenous influences than is the other.

The adaptive flexibility of the bees' spatial orientation was also emphasized by the results of Renner's study in Long Island and California. On the first day in Davis, when the bees searched on New York time, they flew in a direction corresponding to the solar position in New York at that time. That position varied by 45 degrees with the actual azimuth of the sun in Davis, but that variation did not disturb the bees' orientation. By the third day of observation, there was only a slight hint that the bees were switching, during the periods of collection, to using the angle of azimuth at Davis rather than the one in New York to which they had been trained. Would the workers have shifted completely to the California solar time-azimuth relationship, comparable to that of training, had they been observed for a longer period in California?

The results which have been discussed are based on an experiment which was performed only once. Consequently, these results and the interpretations of them should be viewed with healthy skepticism, even though they correlate well with facts of bees' orientation which have been gleaned from many other observations and experiments. The translocation experiment, although costly, deserves repetition.

Another study of the chronometry of *Apis* which should be repeated and extended is one made by us (Bennett and Renner, 1963) during the winter and spring of 1961. We maintained a hive of bees under constant light, temperature and humidity in a bee laboratory in the Institute of Zoology in Munich. We wondered whether workers under those conditions would collect continuously from a constant supply of sugar water. If so, would the level of the collecting activity vary? Would it vary rhythmically? Would there be evidence of circadian cycles? What would the free-running period of a possible cycle be?

We placed the sugar solution inside an automatic recording collecting chamber and trained workers to search there. The bees lived up to their reputation of being busy by continuing to collect day after day. In one case, we were able to record their activity for 40 consecutive days, and in a second run, we were able to gather data for 30 consecutive days. During very few hours of those periods did no bee collect. In fact, the average number of visits to the sugar water was just less than 100 per hour, and more than 2 cc of solution were carried away to the hive every hour. The workers certainly were active, and from observations of the colony and the hive itself, we knew that it was in a healthy state even under the laboratory conditions.

Cycles of a circadian nature were described in terms of the

hourly levels of collecting and solar time (Fig. 23*a* and 23*b*). The periods of the cycles ranged from 19 to 29 hours, and averaged 23.8 hours. Beier (1968) reported an average of 23.4 hours for the cycles of bees maintained under similar conditions. Particular

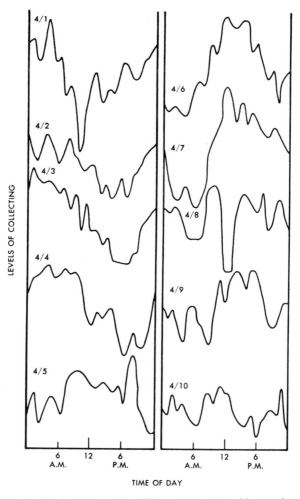

Figure 23.   A. Circadian cycles of collecting activity of honey bees living in a bee laboratory with a constant supply of sugar water. Dates in 1961 are given next to the respective curves. From: Bennett, M.F. and Renner, M.: The collecting performance of honey bees under laboratory conditions. *Biol Bull, 125*:416-430, 1963.

patterns of collecting activity were repeated for two to nine consecutive days. On many of those days, visits to the sugar water were clearly minimal during midday (Fig. 23*a*, 4/1 and 4/8; Fig. 23*b*, 4/26, 5/21 and 5/28).

Noontime sluggishness as indicated by a decrease in flight intensity and a lessening eagerness to dance was also reported for worker bees in their natural environmental conditions (von Frisch, 1967). The laziness could not be correlated with changes in external factors, and was assumed by von Frisch to be an expression of diurnal periodicity. But, a minimum bridging noon was not seen consistently for the bees we kept under constant conditions in our bee laboratory (Fig. 23*a* and 23*b*). In many instances, the forms of the persistent cycles and their phases relative to real time changed, and changed abruptly in a saltatory manner (Fig. 23*a*, 4/1 through 4/5; Fig. 23*b*, 5/21 through 5/29).

How can such "spontaneous" jumps in phasing be explained? Were they caused by unknown and uncontrolled, irregularly recurring Zeitgebers? Were they results of the phases of the cycles of the many worker bees being out of synchrony, and being set and shifted differently? The latter suggestion was made in our original report, and was supported by the following points. We recalled that Beling and later investigators had found that the period of searching at the empty collecting dish could be set for any 1- to 2-hour interval of the solar-day, and that von Frisch proved that nectar laden foragers communicated the presence of food to other workers. The hands of an individual's clock or the time of searching are probably set by the animal's first finding and successfully taking sugar water.

With a group of 40 to 50 bees collecting and communicating with hive mates, is it not further probable that the daily cycles we recorded were the composites of the rhythms of many different workers? For the days when the amplitudes of the rhythms were greater than average and the forms of the cycles were sharp (*e.g.*, Fig. 23*a*, 4/4 and 4/6; Fig. 23*b*, 4/29 and 5/27), we suggested that the rhythms of most of the workers were in synchrony. For those days when the forms of the activity curves

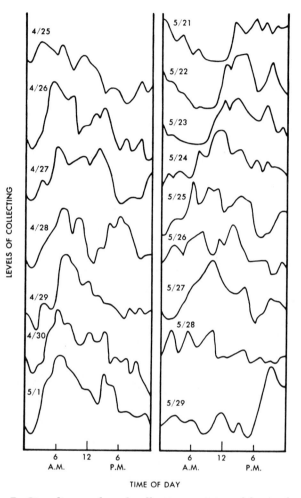

Figure 23. *B.* Circadian cycles of collecting activity of honey bees living in a bee laboratory with a constant supply of sugar water. Dates in 1961 are given next to the respective curves. From: Bennett, M.F. and Renner, M.: The collecting performance of honey bees under laboratory conditions. *Biol Bull, 125*:416-430, 1963.

were less distinct and the amplitudes low (*e.g.,* Fig. 23a, 4/10; Fig. 23b, 5/28), we theorized that the phases of the cycles of the active collectors were not in synchrony. Did the dominance of specific workers increase and then decrease, and thereby affect the forms of the cycles? Was the death of workers whose phases had been set at particular times reflected in irregular shifts in the rhythms as new foragers whose indicators had been set to other points in the solar-day took over the major portion of the collecting? These are all possibilities.

Whatever the basic cause of the changes which were observed, phase-shifting of the individuals' and the group's rhythms must be explained, and the Zeitgebers, or setting agents must be found. The most obvious candidates are the differences in light intensity to which workers are exposed as they fly from the dark hive through the brightly lighted laboratory, into the dark collecting chamber and then return to the hive, because light-dark changes have been identified as some of the most powerful entraining agents of persistent biological cycles. It is staggering to think of the great number of such perturbations to which a bee is exposed when one considers how many times during one round trip she experiences light to dark changes, and how many trips she flies each day.

Also, quantitatively speaking, each perturbation is of great magnitude—from darkness to light of 1000 lux or vice versa. If the workers were exposed to constant light of such great intensity, one might postulate that it evoked some of the changes in period which we observed. Aschoff's Rule (Aschoff, 1960) states that light active animals show a shorter circadian period in constant light than in constant darkness, and that the period decreases with increasing light intensity. Honey bees are certainly light active. Were the periods of only 19 hours the results of the high light intensity even though it was interrupted during the bees' flights? If so, how does one explain periods as long as 29 hours?

Renner and I were able to prove that alternating periods of darkness (12 hours) and of the bright light (12 hours) do set the phases of the cycle of collecting by a group of workers. Under constant conditions, the form and frequency of the rhythms so entrained were maintained for several days (Fig. 24, 6/10 through 6/14), but later the cycle became less precise (Fig. 24, 6/16) and showed saltatory changes (Fig. 24, 6/17). Therefore, the ques-

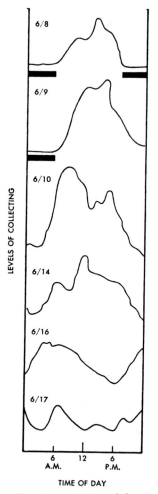

LEVELS OF COLLECTING

6/8

6/9

6/10

6/14

6/16

6/17

6
A.M.

12

6
P.M.

TIME OF DAY

Figure 24. Cycles of collecting activity of honey bees living in a bee laboratory with a constant supply of sugar water during and after exposure to L/D 12/12. The black bars show the periods during which the room was dark, a situation true also of 6:00 PM to midnight on June 7. Dates were all in 1961. From: Bennett, M.F. and Renner, M.: The collecting performance of honey bees under laboratory conditions. *Biol Bull, 125:*416-430, 1963.

tions regarding the contributions of individuals and perturbing factors must be raised once more. The honey bee may present us with a model system for the analyses of individual and population aspects of circadian rhythmicity.

The methods which Renner and I used for our investigations could be modified to eliminate the light-dark changes. One could build glass-walled observation hives and collecting chambers. Every few days, the foragers could be marked. Then they would be known individually, and the activity in the collecting chamber could be recorded constantly on moving picture film. From these films, the investigator could work out the minute-to-minute performances of all individuals and could compare each with another and with the composite of them. Photographic recording of the behavior of the bees within the hive is also necessary to answer the question: are the phases of the collectors' cycles set by dancing or other types of social interaction?

The results of analyses of that type of record might also support the hypothesis of social Zeitgebers developed by Medugorac and Lindauer (1967). Using a mixed group of narcotized and normal bees, they found both types of workers participating in two periods of searching at the collecting table. That was the pattern described earlier for the bees treated with CO_2. The normal bees had been trained to collect only at the time of the first peak of the narcotized bees, yet, the normal workers searched during both intervals. The explanation given was that foragers imprint their times of intense activity on their hive mates. How? Here is another question to be addressed to the honey bee. The answer may teach us more about social Zeitgebers whose existence has been proven in but a few species. The circadian cycles of sparrows (Menaker and Eskin, 1966) and finches (Gwinner, 1966) can be entrained by sounds, those of the species' own songs. As Zeitgebers, social cues seem to be more important than even light changes in human beings (Aschoff, 1971).

Undoubtedly effected by the same cellular mechanism which allows *Apis* to "remember" collecting times at circadian frequencies, is a phenomenon which Koltermann (1969) has called the 24-hour "efficiency" of bees in remembering scents. He trained animals to different odors, and then tested their remembrance of the materials at different times after training. A greater percentage of correct choices of the training odor over other scents was made at 24 hours than at 22 or 26 hours, at 48 hours than at

47 or 49 hours, at 72 hours than at 71 or 73 hours. The results of testing at 1, 3, 6, 10, 18, 20, 24, 28 and 30 hours after training also emphasized the great capacity of remembering at the 24-hour interval. A later study (Koltermann, 1971) showed that many scent training periods, as long as they are separated from each other by at least 20 minutes, are also "remembered" at a 24-hour frequency, and bees can remember training periods to two different odors if these periods are separated by six hours. Lastly, the workers can learn to respond to different colors at circadian intervals.

Which cues might one assume *Specodogastra texana* uses in setting phases of the hands of its clock? This is a solitary bee that forages during twilight and into moonlight hours if moonlight follows twilight without intervening darkness. Moonlight follows twilight directly, at least when the weather is not stormy, on the night of the day after new moon through those of three days after full moon. Kerfoot (1967) described a lunar-monthly cycle of pollen collecting for those bees under field conditions. He presented evidence from the trapping of animals and from dissection of their nests for cycles of 29 days in length. What would the frequencies of rhythms of foraging or collecting of that species be if we were able to maintain individuals under constant laboratory conditions?

When one notes so beautiful a temporal correlation between the behavior of an animal and phases of the moon, he wonders, immediately, about the details of the organism's chronometry. We might well ask of that bee some of the questions we have asked of fiddler crabs. Is a lunar-tidal cycle more pronounced than a circadian one? Are both present? Does their simultaneous functioning yield the lunar-monthly rhythm? We might, in addition, question bees about annual cycles. Bees of the temperate zones face and adjust to the seasonal variations of their environments. A winter hive is a very different place from a summer one (Lindauer, 1971). Do the bees' inner clocks aid in their adjustments to these variations? Bräuninger (1964) described an annual rhythm of variations in the frequency of "waggle" dances of *Apis* which correlated well with the temperature outside the hive. He assumed a cause-effect relationship, but does a persistent seasonal cycle of the bees underlie it?

Numerous other questions about the timekeeping of bees are unanswered. The indicators which are known — time-compensated orientation, the "remembering" of times of training to sugar water, odors and colors and rhythms of collecting activity— are circadian, and their adaptiveness to animals which "live by the sun" do impress us. Are there additional indicators, some of which may be more valuable for the solution of the problems which remain? Which set of hands of the bee's clock can be used most effectively in analyzing the contributions of individuals to the rhythm traced out by a group of workers? Which is best for studies of social Zeitgebers? Which may help us learn most about the cellular clockworks of *Apis*? Which is best for our future analyses of the mediating pathways of the bees' "Innere Uhr."

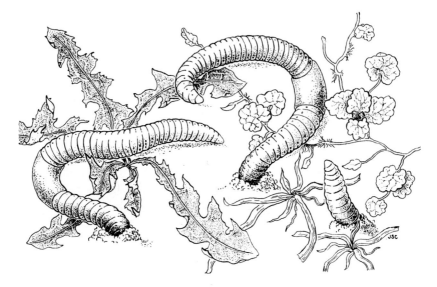

Chapter 5

EARTHWORMS

CHARLES DARWIN'S INSOMNIA is responsible for our first reports of persistent cycles of activity in a common earthworm, *Lumbricus terrestris*, a species which has taught us much about the ways of animal clocks. That creature, as another of its common names, nightcrawler, suggests, comes out of its burrow and crawls about on the surface of the earth during nighttime hours. Daytime for *Lumbricus* is a period of lesser activity spent underground. In his library at Down House, Darwin kept earthworms in pots of soil, shielded from changes in light intensity. He observed their actions periodically, and since he was often unable to sleep, he made detailed observations of the worms during the night when he saw them leaving their burrows. He reported the worms' typical nocturnal performance and the absence of such great activity during the daytime in the introduction to his book, *The Formation of Vegetable Mould Through the Action of Worms,* published in 1881, just a year before his death.

However, Darwin, a student of biological adaptations, did not comment upon the obvious adaptive nature of the cycle of earthworm behavior which he had observed so closely. *Lumbricus* has no exoskeleton or body covering. It faces its physical surroundings naked, as do most of its relatives, the segmented worms of the phylum, Annelida. The annelids which live in terrestrial habitats, notably the earthworms, constantly fight the dangers of desiccation. Body water moves through their thin skins rapidly, especially when the animals are exposed directly to the warm, dry air of sunny daylight hours. Fortunately, the earthworms' timing mechanism tends to keep them in the damp earth at those times, and to allow them to be in the air only at night when the humidity is generally higher and the temperature generally lower than during the day. Their danger of drying up is ameliorated by the working of their circadian clocks.

The possession of a living timer of an annual frequency—an animal calendar—would also seem to be adaptive to the ways of life of earthworms. Several close relatives of *Lumbricus, Allolobophora longa* and *Eisenia rosea,* function very obviously in time with the seasons of the year. During the summer, these worms stay down in the ground in a state of suspended animation or summer sleep called aestivation; during the remainder of the year, they, too, come to the surface of the soil and move about at night. *Lumbricus terrestris,* itself, does not aestivate. On the contrary, in the temperate zones, late spring through summer is this worm's mating period which follows the early spring maturation of its reproductive organs and behavior. One need only observe his lawns on damp summer evenings to see mating pairs of nightcrawlers. Such activities are not to be seen in the fall, winter or early spring. Even though *Lumbricus* does not avoid direct exposure to the summer atmosphere completely, the animal moves into that air primarily when the moisture content is likely to be near maximum. Dangerously high rates of water loss from the body are thus avoided during the mating season, again because of the worms' circadian pattern of locomotor activity which impressed Darwin in the course of his own nighttime wanderings.

Do earthworms have persistent annual rhythms which help keep them in synchrony with the seasons of the year, thus preparing them for aestivation or for annual breeding periods? We have no information about persistent cycles in the aestivating forms, *Allolobophora* or *Eisenia*. Later in this chapter, the details of a yearly rhythm of *Lumbricus* are discussed. The annual cycle insures that these earthworms are able to move especially rapidly on summer evenings when they must crawl toward the surface of the earth to breed. So, the living chronometers of our common nightcrawlers not only equip the creatures to live safely day-by-day, but also season-by-season.

Thirty-five years after Darwin's death, another English investigator, Baldwin, who spent 48-hour stretches watching earthworms, also described their circadian or daily performances (Baldwin, 1917). The animals were exposed to alternating 12-hour periods of light and dark while the observer scored their movements of crawling, feeding and egestion. The worms were three times more active between 6:00 PM and midnight than they were during the other quarters of the day. Similar findings were reported by Szymanski (1918), in whose Viennese laboratory the fiddler crabs, imported from Woods Hole, were studied (Chapter 2). Szymanski automatically and continuously recorded the movements of earthworms exposed to the normal light and temperatures of Vienna in September and October. Their records of crawling showed peaks of activity, each 30 to 60 minutes in length, occurring in the late afternoon and early evening.

The results of experiments on the worms suggested to Szymanski that the periodicity of the animals' locomotion did not demand the presence of all parts of their bodies, for after transection, anterior and posterior halves continued to move like the intact worms had. As is to be discussed later in this chapter, our investigations of the mediating pathways of two different sets of hands of the earthworm clock contradict Szymanski's idea. We are convinced that an intact central nervous system is necessary for the regulation of most circadian variations in the functioning of earthworms.

The daily cycles of the intact worms described by Szymanski and those observed by Baldwin did parallel the rhythm commented upon by Darwin. But, it is impossible to decide whether the periodicities reported by the former two investigators were examples of persistent rhythms or not. Recall that the worms which they studied were exposed to light-dark changes in the laboratories. Those changes may have evoked the different levels of activity rather than entraining phases of actual persistent cycles. Since Darwin's earthworms lived in constant darkness several weeks at a time, his descriptions of their intense crawling at night and quiescence by day constitute more convincing evidence of persistent circadian rhythms in earthworms.

And, the results of a much longer and more detailed investigation of *Lumbricus terrestris* under constant laboratory conditions of light intensity, temperature and humidity increase the validity of the idea of the persistence of cycles in this species. In 1957, seventy-five years after Darwin's death, Ralph (1957a) published his long-term observations of oxygen-consumption and locomotion in earthworms. The animals lived for as long as 30 days in the chambers of Ralph's recording instruments. The simple arithmetical methods, designed by Laplace and described in Chapter 3, were used to analyze the earthworms' rhythms. Among the average cycles of metabolism and locomotion which Ralph found were some of circadian frequency. Generally high levels of oxygen-consumption were characteristic of late morning and early evening, while the amount of crawling was minimal during morning hours and great at night. As far as essentials are concerned, the rhythms reported by Darwin and by Ralph were very much the same.

A provocative study which also revealed the circadian organization of *Lumbricus* was published in 1957 by Arbit, who had investigated the rates at which worms learn to turn in a T-shaped maze. If an animal made the "correct" choice of turns through Arbit's test apparatus, it moved into damp moss. An "incorrect" choice not only took an animal into a tube lined with sandpaper, but the worm was also shocked electrically when it arrived there. The earthworms needed only 32 trials to learn the maze between 8:00 PM and midnight. They required 45 runs to acquire the proper turning tendency when they were trained between 8:00 AM and noon. Two other reactions of nightcrawlers are consistently fast in the evening even when the animals live in the dark and at a constant temperature and humidity. The worms crawl up a 10 cm-long track and jerk away from a spot of bright light approximately 25 per cent more quickly between 7:00 PM and 9:00 PM than between noon and 1:00 PM. These two indicators of the earthworm's solar clock, the rate of crawling and the rate of light-withdrawal, have amused, fascinated and frustrated my students and me since we started our analyses of the rhythms of *Lumbricus terrestris* years ago.

We were looking for a simple, clear-cut reaction which varied predictably with a circadian frequency in an animal that was plentiful and cheap and could be maintained easily in inland laboratories. We planned to use such a system for a functional analysis of the many components of an organism's timing mechanism. After only several weeks of observation we found that the light-withdrawal reflex of earthworms fit our specifications. That reaction can be demonstrated readily. If a bright light is shined on the anterior end of a worm as it crawls along, the animal jerks back, and consequently usually removes its body from the intense illumination. Similarly, if a light is shined on the worm's tail or on the middle portion of its body, the part illuminated is rapidly withdrawn from the light. One who is interested in the circadian variations in the speed of this reflex need only time the worms' withdrawal around midday and again early in the evening. An analysis of several days' data shows that the animals are on the average 20 to 25 per cent faster at night (Bennett and Reinschmidt, 1965a).

To eliminate as many variables as possible in our investigations of the light-withdrawal reaction, we have always purchased the worms, *Lumbricus terrestris*, from the same supply house, and have kept them in the laboratory under identical sets of conditions. Each worm lives in a glass container with damp loam in the darkness at a cool (17°C) and constant temperature. Only at the times of testing are the worms removed from the incubators in which the constant conditions are maintained. Then they are taken to laboratories illuminated diffusely by dim (less than 1 lux) red light. When a worm is tested, it is placed on its ventral surface on a black bench. As it crawls along, the investigator shines a circle of bright white light (10 lux) on the animal's head, and determines the time necessary for the worm to leave the lighted area. Three of my beginning students have proved that yellow, blue and green lights are just as effective as is white in evoking rapid light-withdrawal in the worms.

There certainly are individual variations in the actions of earthworms. Now and then, one may react more slowly in the evening than it had at noon. However, over a 2- to 3-week period, a group of five to ten animals withdraws from the light on the average 20 to 30 per cent faster at night than at midday. To guard against my biasing the results, I have hired undergraduates, who were unaware of earlier studies, to test worms for me. They have consistently recorded data comparable to mine. In addition, the students have commented upon the great difference in the manner in which earthworms perform at the two times of day. During midday tests, the worms are said to move without precision or like "soft spaghetti." In the evening, their reactions are snappy and directed.

Our first studies of the light-withdrawal reaction ran through late winter and early spring (Bennett and Reinschmidt, 1965a). The results helped us set the background for many of the analyses of the clock and the calendar of earthworms which have succeeded that first study. The worms we tested in February and March reacted 31 per cent faster around 8:00 PM than at midday. The group we studied in April and May was only 24 per cent more rapid at night. Consequently, we knew that an obvious circadian difference in the rates of the reaction existed, and we had an idea of the magnitude of that difference. The variation in the magnitude of the circadian difference between late winter and early spring hinted that lunar and seasonal cycles might also influence the actions of earthworms.

Our findings also raised a number of questions concerning the mechanism and the regulation of the timing phenomenon

which causes *Lumbricus* to pull away from light so much faster
in the evening than around noon. The worm's ability to move
speedily at night—when it is likely to be abroad in its natural
environment—is beautifully adaptive to its way of life, as was
stressed earlier in this chapter. How can we explain that capacity? Do the light receptors, specialized cells in the body walls
of earthworms, function at a greater rate in the evening than at
midday? Do the muscles of the worms contract and relax faster
at night than during the daytime? Do their nerve cells and nerves
act at maximal speeds during the evening?

Any one of these propositions could explain the solar variations in the light-withdrawal reflex. If all three suggestions were
fact, the reaction would indeed occur faster at night than at
noon. Conversely, if no one of these types of cells, receptor,
muscle or nerve, were independently rhythmic on a daily
schedule, the withdrawal reflex could still be faster in the evening
than around noon. To explain that situation, one could postulate
that the worms' general, cellular clockworks, the specific location
and function of which are unknown, forced their circadian oscillations on one or on all the links of the reflex pathway necessary
for the withdrawal from light. If that were the case, hormones
or nerve impulses might well be the mediators between the basic
clocks and this indicator of the rhythmicity, the rate at which
Lumbricus jerks itself out of the light.

To start our analysis of the timing of light-withdrawal, we
tried to answer this question, *viz.*, whether or not it is stimulated
by light, does *Lumbricus* move faster in the evening than at
noon? We reasoned that if the answer were *yes*, we might be
able to eliminate the actions of the photoreceptors and their
nerves as the basic timers or pacers of the circadian variations
established for the withdrawal reflex. Crawling during different
periods of the solar-day was timed. The worms were found to
move at greater rates early in the morning and during the usual
evening test period than they did around midday (Bennett and
Reinschmidt, 1965b). We not only had the hint that the photoreceptors were not necessary for the *timing* of the reflex action,
but also had identified another indicator of the circadian or-

ganization of earthworms, the rate at which they crawl. That is certainly a reflection of the rate at which their locomotor muscles function. What regulates the speed of the muscular actions? Our answer to that question is to be presented later in this chapter.

The earthworms which we used for our studies of crawling or locomotion came from the same stock from which we secured the worms whose light-withdrawal was investigated. All the worms lived under the same constant laboratory conditions between test periods. Crawling was timed in the morning, around midday and between 8:00 PM and 9:00 PM in the same laboratories in which the light-withdrawal was timed. We determined the time necessary for earthworms to move 10 cm up a wooden ramp set at 45 degrees with the horizontal. The worms were forced to move in straight lines for the surface of the ramp was grooved. Consequently, finding their speeds of locomotion was a simple procedure.

Our first detailed investigation of locomotion as an indicator of the circadian organization of *Lumbricus* was run through the fall and winter. As has been mentioned, the earthworms were slower at noon than in the morning or evening (Fig. 25). The average differences between the rates of crawling were greater than 20 per cent. The temporal variations in locomotion paralleled those of the light-withdrawal reaction. Obviously, the worms' photoreceptors need not vary through the day to account for the differences in rates of their locomotion. Only changes in the movements of their body walls—however they are induced or regulated, can explain the daily differences in crawling.

Our early experiences with both the indicators of solar rhythmicity, the withdrawal reaction and locomotion, caused us to suspect that there were longer-term cycles superimposed upon the circadian ones. For three years, using the rate of crawling as an indicator, we looked for the possibility that earthworms function at lunar and annual frequencies. On as many days as possible, we timed locomotion at midday and in the evening, and calculated the monthly averages of the differences between the rates of crawling at the two times of the solar-day.

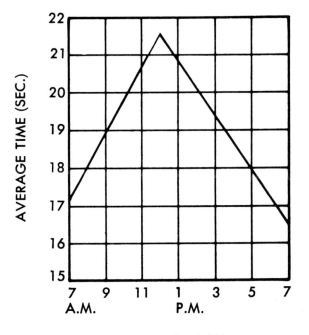

TIME OF DAY

Figure 25.   The average number of seconds necessary for earthworms to crawl 10 cm around 7:00 AM, noon and 7:00 PM. From: Bennett, M.F. and Reinschmidt, D.C.: The diurnal cycle and locomotion in earthworms. Z *Vergl Physiol, 51*: 224-226, 1965b., Berlin-Heidelberg-New York, Springer.

An annual cycle of activity was established (Bennett, 1968), and is presented in Figure 26 where one can see its very obvious general features. During all months of the year, the worms crawl faster in the evening than at noon. However, the magnitude of the difference between the rates of locomotion at the two times of the solar-day varies with the season of the year. A low of only 18 per cent solar variation was found for January while a high of 47 per cent was calculated for August. The annual cycle of *Lumbricus* is a simple one, with a gradual increase in the circadian differences following the minimum of winter and continuing until the maximum of late summer. After the peak in August, the size of the solar variations decreases and remains much the same through the fall and winter.

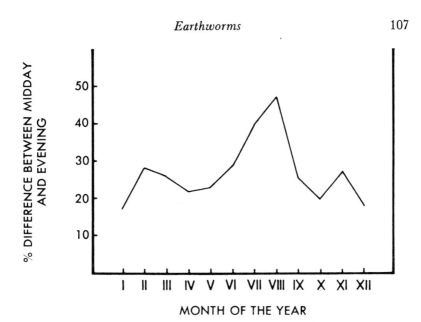

Figure 26. The average per cent differences between the times necessary for worms to crawl 10 cm at noon and in the evening during the year. From: Bennett, M.F.: Persistent seasonal variations in the diurnal cycle of earthworms. *Z Vergl Physiol, 60*: 34-40, 1968, Berlin-Heidelberg-New York: Springer.

A persistent rhythm with these characteristics does seem adaptive to *Lumbricus terrestris*, for, as noted earlier in this chapter, this species has a definite yearly reproductive cycle. Its gonads and other sex organs mature through the spring and early summer, and then regress after the mating period of high summer. Its persistent annual rhythm endows the nightcrawler with the ability to move most rapidly during the nighttime hours of summer when it is ready to mate and wanders about near the surface of the ground.

The earthworms' elegantly uncomplicated annual cycle is a valuable one with which to try to analyze persistent yearly rhythms in general. Descriptions of seasonal cycles of organisms which are maintained under constant conditions of light, temperature and humidity are much rarer than are reports of persistent cycles of solar or lunar frequencies. The major reason for the situation is not difficult to find. An almost staggering number of data, collected over several years' duration, is required to identify cycles of close to 365 days in length. Nevertheless, some of us have made such collections, and have been able to verify that in addition to *Homo sapiens,* earthworms, potatoes, algae, wheat seedlings, migratory birds, crayfish, frogs, rats and ground squirrels live by calendars (For review, see Bennett, 1968). Granted, the calendars may not all be academic, let alone Gregorian, but they can be deciphered by *Homo* if he spends two or three of his years observing and recording appropriate activities, *e.g.,* metabolism, growth, nuptial and reproductive behaviors, of other species which can live in his laboratories.

The annual cycles which have been reported to date share some important primary features (Bennett, 1968 and Chapter 7). They are usually of very simple form with one broad peak and one broad low. In addition, the great differences in the levels of the indicator processes occur between two roughly circumscribed parts of the year, spring-summer and autumn-winter. *Lumbricus* moves relatively much faster on summer evenings than on winter ones; potatoes respire at higher rates in the spring and summer than in the fall and winter; the livers of rats store more carbohydrate in September than in March; ground squirrels hibernate in the winter and are wakeful for the remainder of the year.

Mathematical correlations between phases of annual cycles of organisms and of abiotic factors are cited often by Brown (1960, 1965 and 1970) as evidence supporting the theory that the frequencies of cellular timing mechanisms are generated with the help of subtle geophysical factors. Common vegetables have provided much of that information. The respiratory rates of plugs of potato tubers, pieces of carrots and germinating bean seeds have been recorded at a *constant* temperature, day after day, month after month. Yet, the amount of oxygen used by the plants paralleled the average outdoor temperatures. Oxygen-consumption was found to be maximal in the summer and minimal in the winter. The organisms also lived under *constant* barometric pressure while in Brown's respirometers (Chapter 3), and correlations between the rates of respiration of potato tubers and the barometric pressure changes, recorded in the laboratory, were numerous.

We ask again: what geophysical factors affect the barometric pressure and temperature of the atmosphere and aid in keeping biological clocks and calendars in synchrony with cycles of our physical world? Students of biochronometry who are convinced that cellular timing mechanisms are completely endogenous do not even ask that question. They account for circannual rhythms as they explain organismic cycles of solar and lunar frequencies. Cellular clockworks have the inherited and inherent capacity to oscillate independently of their environments with all those period lengths: 24 hours, 12.4 hours, 24.8 hours, 29.75 days and 365 days.

To date, no one has actually proved whether a yearly or any other cellular timer is partially exogenous or completely endogenous. We must wait for proof, but in the meantime, there is much to be learned about persistent annual cycles. Do their forms and amplitudes vary from year to year? How easily can their phases be shifted? What are their common Zeitgebers? How do they respond to various perturbations? The seasonal variations in the locomotory movements of *Lumbricus* provide us with a timing system from which we can learn the answers to some of these questions. Perhaps there is also an annual fluctuation in the rates of the nightcrawlers' light-withdrawal reflex which would be useful in our analyses.

The variations in the degree of difference between midday and evening crawling and between midday and evening light-withdrawal may also be valuable in our attempt to analyze lunar cycles. Surprising as it seems, there is evidence that terrestrial as well as fresh water plants and animals live in time with phases of the moon and hence with phases of the "tides of the atmosphere." Potatoes, salamanders, some insects, rats, mice and man, himself, are known to function at 24.8- hour and 29.75-day frequencies (Brown, 1970 and Chapter 7). Average cycles of lunar periods were also described for *Lumbricus* by Ralph (1957a), who used locomotion and oxygen-consumption as indicators of earthworm clocks. Unfortunately, there was little consistency between the phasing of the cycles of the two activities, and the forms of the average lunar rhythms varied greatly from semilunar period to semilunar period. From Ralph's analyses one is not convinced that earthworms live by the phase of the moon.

Likewise, relationships between phases of the lunar-monthly cycle and the rates of crawling in *Lumbricus* merely hint at the possibility of 29.75-day rhythms in this worm (Bennett, 1968). From December through May, the per cent differences between midday and evening speeds of locomotion are considerably greater for semilunar periods centered on the day of full moon than for 15-day periods centered on new moon. However, that correlation does not hold for the remaining months of the year. Do nightcrawlers really function at lunar frequencies, or are our hints based on nothing other than numerical coincidences?

The question should be answered, for if *Lumbricus,* a well distributed terrestrial form, possesses lunar chronometers, a very neat system with which to study the possible adaptive nature of living in time with the moon in a nontidal environment is available. Perhaps the lunar rhythms of fresh water and terrestrial organisms do not confer any advantages on their possessors. Perhaps these rhythms are nothing but behavioral relics inherited from marine ancestors whose lives depended upon their synchronizing vital activities with phases of the moon and phases of the tides. Very possibly, students of the evolution of behavior would find the story of earthworm clocks and calendars intriguing and valuable for their investigations.

My students and I certainly have found that story invaluable, for it has provided the background for our investigations of the mediation or regulation of the hands of animals' clocks. Early in our experimental attack on that problem we were able to conclude that temporally varying functions of earthworms' photoreceptors and the optic nerves did not necessarily explain the circadian differences in the actions of our worms. Nevertheless, as has been emphasized, the light-withdrawal reflex is a sharp indicator of solar-day variations, so we have used it and the speed of locomotion in our attempt to explain the mechanics of timing in *Lumbricus.* Our first question was: is the brain, or more properly, are the suprapharyngeal ganglia required for the regulation of the midday-evening variations in the rates of the worms' movements? Our answer is yes (Bennett and Willis, 1966).

Mary Willis Finlay and I followed a very simple procedure in securing that answer. We removed the brains from our experimental worms, and performed sham operations, just cutting the body wall over the brain on control animals, and then compared the rates of actions in the two groups of *Lumbricus*. One can quickly and easily remove the brain from an earthworm (Fig. 27). The thin body wall lying over the brain is slit. Then the circumpharyngeal connectives or large nerves which run from the sides of the brain down to the pair of subpharyngeal ganglia which lies ventral to the gut are cut. Next, the brain is lifted and removed from its position dorsal to the digestive tube. The wounds heal within a day or so, after which time we began the testing of our worms.

Most worms which have lost their brains live just as well as normal earthworms do under laboratory conditions. Generally, brainless *Lumbricus* are restless and crawl with their heads up in the air, but according to many zoologists who have observed

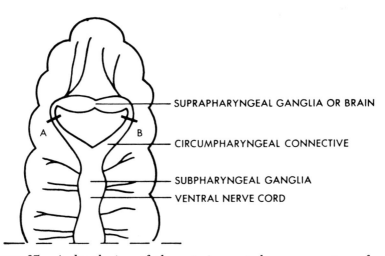

SUPRAPHARYNGEAL GANGLIA OR BRAIN

CIRCUMPHARYNGEAL CONNECTIVE

SUBPHARYNGEAL GANGLIA
VENTRAL NERVE CORD

Figure 27. A dorsal view of the anterior central nervous system of an earthworm and the cuts (A and B) made when the brain is removed or is separated from the more posterior parts of the nervous system.

their behavior, such worms can right themselves, can burrow into the soil and can learn simple mazes.

However, some part of their circadian organization is awry after the loss of their brains. We kept groups of brainless and sham-operated worms under the same constant laboratory conditions, and timed their light-withdrawal and locomotion during the usual noon and evening test periods for two weeks following the operations. The control earthworms were approximately 25 per cent slower around midday than at night. The brainless animals were just as fast at noon as in the evening. The typical solar-day difference in the rates of the two indicator processes had disappeared (Bennett and Willis, 1966). Might that circadian variation reappear in worms which had regenerated their brains? Regeneration of the suprapharyngeal ganglia of *Lumbricus* has been reported many times, and investigators of that phenomenon may consider our question particularly interesting. We have never kept our experimental earthworms long enough after their simple surgery to provide an answer.

Later work done in my laboratory (Bennett, 1967 and 1970) has provided a partial answer to another question. How does the brain or the pair of suprapharyngeal ganglia regulate the timing of earthworms' actions? That organ exerts its effects as a classical and typical neural structure via bioelectrical changes, probably nerve impulses. Conversely, neurosecretions, or chemical regulators secreted by some cells of the brain, are not primary mediators of the solar variations in the rates of the worms' light-withdrawal and crawling. Chemical mediation of timing, comparable to that known so well for rhythmic color changes in fiddler crabs (Chapter 2), was held possible for the nightcrawler. Neurosecretory cells are present in the brain of *Lumbricus* as they are in the brains of other annelids, marine and fresh water ones.

The late Ernst Scharrer, whose pioneer observations and analyses of neurosecretory structures and functions led to the development of our present, sophisticated ideas of neuroendocrinology, published electron photomicrographs of neurosecretory cells in the suprapharyngeal ganglia of earthworms (Scharrer and Brown, 1961). The products of these cells are believed to regulate body water content and reproductive functions of the worms.

But, the results of my experiments on timing do not suggest the products of those neurosecretory cells as the regulators of persistent rhythmicity in nightcrawlers. I timed light-withdrawal and locomotion, once again around midday and in the evening, in sham-operated animals and in worms in which I had severed the connectives between the brain and the subpharyngeal ganglia (Fig. 27). The experimental earthworms were treated very much as our brainless worms had been, except that after cutting the connectives, the brains of the new series of experimentals were left in place on the dorsal surface of the gut. It was assumed that if neurosecretions of the suprapharyngeal ganglia were the major regulators of the rates of the actions being studied, their influence would continue even after the large nerves leading from the brain had been cut.

Neurosecretions are released ultimately to body fluids in which they are carried to all parts of animals' bodies, so those produced by an isolated brain might still communicate normally with their targets, the cells, tissues or organs typically regulated by chemical messengers. On the other hand, if the mediation of the circadian variations in rates of reactions in *Lumbricus* were effected by nerve impulses traveling in specific pathways between the brain and the more posterior parts of the nervous system, the severing of the circumpharyngeal connectives, which are the major pathways themselves, could be assumed to disrupt that regulation. I believe that is the explanation for the results which I recorded. The worms with brains that had been isolated by the cutting of the large connectives jerked out of a spot of light and crawled at the same speeds during the two test periods. Once again, I found that the circadian variation had disappeared (Bennett, 1967).

Does the regulation of the normal daily variation in the rates of earthworms' movements require information passing between both sides of the brain and other parts of the nervous system? The bilateral symmetry of the brain is apparent, grossly, and there is a pair of connectives between the paired supra- and subpharyngeal ganglia (Fig. 27). Not long ago, I was able to answer the question posed above. In the experimental worms, I cut either the right or the left member of the pair of circumpharyngeal connectives. (I made cut A or cut B illustrated in Fig. 27). As is usual, I merely slit the skin lying over the brain in the control animals. The controls withdrew from light and crawled 20 per cent faster at night than at noon. The experimentals, whether the right or the left connective had been severed, reacted to light and moved at approximately the same speeds during the two test periods of each solar-day (Bennett, 1970).

A completely intact anterior nervous system is demanded for the timing of those two indicator processes. A probable explanation is that output from only one side of the brain is not great enough to transmit to the posterior parts of the nervous system all the information required for varying the rates of light-withdrawal and locomotion at various phases of the solar cycle.

I continue to believe that such information is carried primarily as nerve impulses rather than by neurosecretions of the brain. Nevertheless, there is another possible explanation for the disappearance of the circadian differences in *Lumbricus* after the cutting of one or both of the pathways between the brain and the posterior nervous system, and the explanation is based on neurosecretions as regulators!

Normally, neurosecretory cells of the brain might release their products into sinuses or neurohemal organs located in the circumpharyngeal connectives or in the subpharyngeal ganglia. Elementary neurohemal organs have been observed in anterior parts of the central nervous system of relatives of *Lumbricus*. If in severing the connectives of our experimental animals, we isolated the bodies of neurosecretory cells from their axons, the extensions that release the hormones into sinuses which contain some of the body fluids, the secretions might not be moved into the body fluids at all. Or, cutting the circumpharyngeal connectives might destroy neurohemal sinuses themselves. In these cases, it is conceivable that regulation mediated by the neuro-secretions would cease. Earthworm timing would be set awry.

We must become very well acquainted with the ultra-structure of the suprapharyngeal ganglia, the circumpharyngeal connectives and the subpharyngeal ganglia of *Lumbricus* to see whether my propositions deserve serious consideration. In resolving the anatomy of these structures at the level of electron microscopy, we may be able to map out the precise nerve pathways that carry the impulses which are postulated to constitute the regulatory information of earthworm timing. To prove that idea, electrophysiological analyses must also be accomplished.

The splendid recording from single neurons of sea hares, described in Chapter 6, and that done from crayfish nerves, described in Chapter 3, give hope that comparable information can be secured for earthworms and other animals whose indicators of persistent rhythmicity are mediated by neural activities. That type of regulation is primary in the circadian cycles of locomotion in cockroaches (Brady, 1969). On the basis of experimental attacks very much like those we used on *Lumbricus*, investigators are now convinced that nerve impulses carried through particular chains of neurons tie the cellular clockworks of the roaches to their muscles of locomotion. Whether or not signals from both sides of cockroaches' brains are necessary for that regulation remains to be determined.

The finding that both circumpharyngeal connectives of night-crawlers must be intact if the circadian variations in the rates of the worms' movements are to persist illustrates a case in which the dual functioning of bilaterally arranged neural structures is necessary for particular actions. Questions and ideas regarding the roles and the importance of bilaterally arranged nervous structures have also come from other types of investigations (For review, see Bennett, 1970).

Generally, it seems to have been assumed that comparable units are present on both sides of a bilaterally symmetrical nervous system, and that their functioning is very similar if not identical. We know that some patterns of animal orientation are upset when one of a pair of receptors is nonfunctional. Man's three-dimensional vision is greatly impaired by the loss of sight in one eye or by damage to optic areas of one side of his brain. Yet, in most human beings, only units in the left cerebral hemisphere are necessary for normal speech, and only one side of birds' forebrains controls their sound production.

But, in crickets, the members of a particular pair of connectives have been proved to have slightly different functions. Too much has been assumed. We must set about proving whether or not comparable units exist on both sides of a bilateral nervous system. Their functions must be worked out in detail. The adaptive natures of symmetrical and of asymmetrical nervous systems must be considered. *Lumbricus terrestris* may very well help us learn more about a bilaterally symmetrical system.

That worm has already taught us that the workings of its central nervous system in the mediation of timing are complex. The loss of circadian variations in light-withdrawal and locomotion which follows the removal of the brain, the cutting of both circumpharyngeal connectives or the severing of only one of these nerve pathways has been stressed. Is an intact, anterior central nervous system required for the regulation of any other indicators of earthworms' rhythmicity? Here, only an equivocal answer can be provided (Bennett and Guilford, 1971).

A third indicator process identified in my laboratory is the difference in the rates of midday and evening oxygen-consumption in pieces of nightcrawlers. The pieces, used for only one recording session, and consisting of the prostomium and the first six segments of a worm, are cut from animals which have lived under constant laboratory conditions of light, temperature and humidity for several days. The preparations are made just before metabolism is measured. Consequently, we are reasonably certain that the variations recorded have not been entrained by our laboratory procedures. Pieces of the anterior ends of normal worms consume approximately 25 per cent more oxygen between 6:00 PM and 8:00 PM than around noon (Bennett and Guilford, 1971). The magnitude of that variation and the increase in evening over midday rates parallel the characteristics of the solar-day differences in the speeds of light-withdrawal and crawling. But, here, the neat parallelisms among the three indicator processes cease.

Preparations from worms whose brains were removed or isolated by cutting the large connectives between the brain and the more posterior parts of the nervous system, only an hour before their metabolic rates were measured, also used 25 per cent more oxygen during evening hours than during midday ones. The circadian variation in rates of oxygen-consumption did not disappear at least within several hours following surgical manipulation of the anterior central nervous system (Bennett and Guilford, 1971). Does that mean that the timing of rates of earthworm metabolism is not normally accomplished by neural activities?

Perhaps the circadian changes in oxygen-consumption in *Lumbricus* are similar to those which have been recorded for organisms which lack nervous tissue. It has been postulated that the rhythms of respiration in such forms as seaweeds, carrots, bean sprouts and potatoes are very close reflections of the states of the cellular clocks themselves, and do not depend upon a specific mediating tie-up (See review, Bennett, 1973).

By extrapolation, we might suggest that the variations in the levels of oxygen-consumption in the nightcrawlers are closer reflections of the conditions of their cellular timing mechanisms than are the changes in the rates of light-withdrawal and locomotion.

These activities, as has been emphasized repeatedly, indicate cellular timing, but can do so precisely only when many reactions, receptor, integrating and muscular ones, are synchronized by the intact nervous system. In short, one could conclude that no mediating pathway is involved between cellular timing and the levels of oxygen-consumption in earthworms, while one—the anterior nervous system—must operate between the cellular clocks and motor reactions if these reactions are to vary at solar frequencies.

However, we are not convinced that the changes in the levels of metabolism are not normally regulated by the nervous system. The preparations whose oxygen-consumption was measured were cut from worms whose nervous structures had been removed or severed only an hour before the measurements were begun. Therefore, this possibility must be acknowledged: nervous activity does mediate the circadian changes in respiration as it does those of locomotion and light-withdrawal. In the cases of the earthworms whose brains had been removed or isolated shortly before we recorded oxygen-consumption, functioning of the anterior nervous system had already brought the respiratory reactions of all cells close to their typical midday or evening rates. These rates were then maintained for several hours without information from the nervous system. The brainless nightcrawlers and those with isolated brains which have been used in our studies of locomotion and light-withdrawal were tested at earliest 24 hours after their simple surgery; their solar-day differences had disappeared.

Is there any evidence that circadian variations in oxygen-consumption disappear as the time after disruption of the nervous system increases? Yes, there is, but it is ambiguous. We removed the brains from worms roughly 42 hours before the oxygen-consumption of preparations cut from them was recorded either at noon or early in the evening. The average respiratory rate was still higher between 6:00 PM and 8:00 PM than around midday, but that increase was much less—only 16 per cent—than the 25 per cent one found for normal worms (Bennett and Guilford, 1971). Did that indicate that the solar-day difference was damping out? Or, by chance did we study worms that had respiratory cycles of lower than average amplitudes?

These experiments, which are very time-consuming and tedious, are being repeated and extended, for we find it of great importance to be able to compare the details of mediation and regulation of several indicators of rhythmicity in a single species. A clearer understanding and appreciation of the spectrum of types of biological timing phenomena may come from such comparative investigations.

More studies of the effects of physical factors on the many aspects of biochronometry will also help establish the extent of that spectrum. To be noted is the possibility that one end of the spectrum is represented by living clocks that are completely endogenous while the opposite end is represented by organismic timepieces which can not generate solar, lunar or annual frequencies without temporal information from their geophysical environments.

Lumbricus, the earthworm which has taught us about the mediation of solar-day variations, has also contributed information underlining the importance of a physical factor, geomagnetism, to organismic timing. We secured that information in my laboratory using the nightcrawlers' light-withdrawal reaction as an assay of the influence of magnetic force on circadian variations. Control earthworms lived in darkness and at a constant temperature in the earth's normal magnetic field. The experimental animals were kept at the same temperature and also in the dark, but lived and were tested in an artificial field in which the magnetic force was essentially zero.

Through the fall, we timed the reactions of the worms from the two environments. The control animals were 20 per cent faster in the evening than they were around midday. There was no difference between the average withdrawal times of the experimental worms at the two times of day (Bennett and Huguenin, 1969). As after the removal of the brain, the cutting of both connectives or the severing of only one of the circumpharyngeal connectives, the circadian variation in the rate of the light reaction disappeared in worms deprived of normal magnetic flux. That variation was also absent in groups of *Lumbricus* which were maintained and tested in a magnetic field of twice the force of the earth (Bennett and Huguenin, 1971). Geomagnetism certainly affects the circadian organization of *Lumbricus.*

There is good reason to believe that geomagnetism also affects cyclic responses in several unicellular organisms, flatworms, mud snails and man. The investigations of magnetic force and the orientation of mud snails are reviewed in detail in Chapter 6. The results of studies of geomagnetism and its relationships to human rhythms should be considered seriously by those responsible for and interested in man's performance in space. While he is away from the influence of terrestrial rhythms, man could suffer desynchronization of some of his physiological cycles. Bonny, the space monkey scheduled for a 30-day space flight, was brought down on the eighth day of his mission because of physiological abnormalities including desynchronization of his temperature, cardiac and respiratory rhythms. The macaque died from heart failure shortly after he was removed from his capsule.

Rütger Wever, who has monitored temperature, excretion, activity and rest in almost 100 different persons living under controlled conditions in bunkers in Bavaria, has discovered much greater desynchronization in individuals exposed to magnetic fields of close to zero than in those who were in bunkers where magnetism was of natural intensities (Wever, 1967 and 1971). He also has evidence that changes in magnetic force constitute Zeitgebers for human cycles, and that the amplitudes and the average levels of circadian rhythms in man are influenced by geomagnetic flux.

Of course, Bonny, the space monkey, was exposed to more than atypical or abnormal changes in magnetism. A host of geophysical aberrations greets the space traveler from earth. But in view of Wever's observations, should we not provide our astronauts with artificial magnetic fields approximating the strength and alternation of those of the earth? These combined with cycles of light and dark might insure healthier, more efficient and safer space life for us terrestrial organisms.

Students of numerous biological specialties believe magnetic force to have important effects on living systems (For reviews, see: Barnothy, 1964 and 1969 and Presman, 1970). However, ornithologists continue to question the role of geomagnetism in bird orientation and navigation. European robins are claimed to respond to artificial fields (Wiltschko, 1968); Indigo Buntings are said not to respond (Emlen, 1970); homing pigeons can respond to magnetic fields (Keeton, 1972). Deviations in the dances of honey bees (Chapter 4) correlate with the intensity and inclination of the earth's magnetic field (Lindauer and Martin, 1968), while a greatly increased magnetic force inhibits the firing rates of groups of cockroach nerve cells (Russell, 1969). In human beings exposed to an experimental magnetic force more intense than that of their normal environments, simple reflex actions were slowed (Friedman, Becker and Bachman, 1967).

Explanations for the effects of magnetism on living systems are not numerous. Questions abound. Are specific magnetic receptors necessary or does every cell contain molecules which are sensitive to magnetic flux? Does that force affect bioelectrical phenomena primarily? Are enzymes especially prone to magnetic disturbances? Our future analyses of relationships between geomagnetism and long-term organismic cycles may give us hints generally applicable to the many problems concerning magnetism and life. In any event, we must understand the workings of that force on timing mechanisms of living forms if we are ever to have a complete story of biological clocks. Brown (1970) is convinced that geomagnetic flux participates with the cell in generating physiological cycles; Wever (1971) concludes that the force is another Zeitgeber. From our work on geomagnetism and *Lumbricus*, we can only suggest its possible roles.

As has been described, we used the variation in rates of light-withdrawal at two times of the solar-day as a test system of the effects of magnetism on the circadian organization of the earthworms. The atypical forces affected the worms' performances. But we do not know whether the changes in geomagnetism shifted the phases of the animals' cycles, changed the frequency of their rhythms or inhibited the normal functioning of their cellular clockworks. It seems equally plausible that the photoreceptors, the brain, its connectives or the body wall musculature of the experimental worms functioned abnormally when deprived of magnetic flux typical of their natural habitats. Even to approach a firm explanation of our results, the precise form and the free-running period of the cycle of rates of earthworms' light-withdrawal must be determined. We are obtaining such information. It is not easy, for worms which are handled often rapidly become moribund. Dying *Lumbricus* is not a reliable animal clock.

Nevertheless, we have some idea of the daily patterns of light-withdrawal and of the speed of crawling of nightcrawlers (Bennett, 1969). The curves presented in Figure 28 and Figure

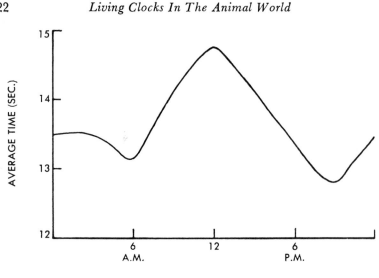

Figure 28.   A 10-day average solar cycle of the times necessary for earthworms to crawl 10 cm. Timing was done every 3 hours for worms of groups of 10 in June, 1969.

29 are based on the timing of the two activities every three hours through two- to three-day blocks of time. The cycle of the changes in the locomotory rate is a simple one (Fig. 28). The worms moved rapidly from 9:00 PM through the early morning hours. They were generally slower during the daytime, and their minimal rate was observed at midday. The rhythm of light-withdrawal is a simple one, too (Fig. 29). Again, the worms were fast during the night and early morning, and were slower during the day. Their slowest reactions were recorded at 3:00 PM.

So, as Charles Darwin watched earthworms by day and by night, so have we. That approach, valuable as it has been, is severely limited. We are going to automatic and continuous recording of oxygen-consumption, locomotory activity and electrophysiological changes in *Lumbricus*. From our records, we hope to ascertain the free-running periods of the cycles of nightcrawlers and their activity to rest ratios. Phase-shifting and entrainment can also be investigated with automatic recording. We want to locate the exact nerve centers and pathways involved in

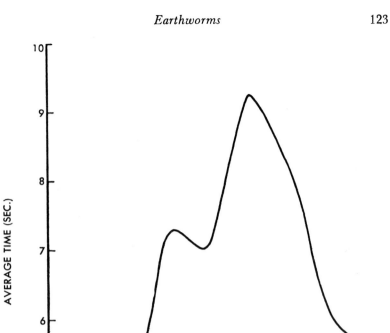

AVERAGE TIME (SEC.)

TIME OF DAY

Figure 29.   A 9-day average solar cycle of the light-withdrawal times of earthworms. Timing was done every 3 hours for animals of groups of 10 in June, 1971.

the regulation of the hands of the clocks of earthworms. *Lumbricus terrestris* is too valuable a form with which to pursue details of animal clocks to leave the questions posed in this chapter unanswered. And, other earthworms, the aestivating species mentioned earlier and the giant ones of Australia, probably also behave rhythmically, so records of their activities should be intriguing. Certainly the clocks of marine annelids, whose spawning correlates so precisely with phases of the moon and of the tides, deserve more attention than they have received.

Persistent rhythms of a few nereid worms have been described. As early as 1904, Bohn commented upon the activity of some of these which he kept in an aquarium in his laboratory. The worms came out of the sand at the times of high tides, and burrowed into the sand during the times of low tides as they occurred in the animals' original environments. The French investigator attributed that tidally correlated behavior to the polychaete worm's "memory" of the impacts of waves in its native habitat.

A more analytical series of studies of cycles of *Platynereis* is coming from the laboratories of Hauenschild in Germany. He has described lunar rhythms of swarming of these worms, and has established that a photoperiodic factor is a normal Zeitgeber (Hauenschild, 1961). However, after worms' eyes were removed, the phases of their cycles could still be set by light. Comparable observations have been made for crayfish (Page and Larimer, 1972) as well as for some amphibians and birds (Chapter 7). Exciting questions about extraoptic photoreceptors and persistent rhythms are discussed in Chapter 7.

Also exciting to the student of annelid clocks are the descriptions of cycles of color change in *Platynereis* (Röseler, 1970). The dispersion and concentration of pigment in three different types of color cells, small red ones, large red ones and white ones, in the skin of this marine worm signal phases of solar-days and of lunar-months. The mediating mechanism of this indicator process is said to be a double one, partially dependent upon nervous activity of the brain and partially dependent upon a chemical factor produced in the posterior part of the ventral nerve cord.

The work of Hauenschild and his associates indicates that characteristics of persistent rhythms and of reproductive behavior are related in *Platynereis* as we have found them to be in *Lumbricus*. Therefore, one is tempted to wager that comparable correlations exist in the marine annelids whose lunar swarming is accompanied by bioluminescent flashing. The fireworms of

Bermuda spawn just after sunset during a few evenings following the day of full moon. The females come up to the surface emitting a greenish glow. The males follow flashing their light signals. When the two sexes swim close to one another, their bodies burst. Their eggs and sperm are shed into the sea as the adult worms die.

Could one maintain these worms in the laboratory long enough to record their movements and their bioluminescence? These activities might be very neat and precise indicators of the clocks of fireworms whose rhythmic displays in their natural environments have fascinated naturalists and explorers for centuries, and whose timing mechanisms may contribute to further understanding of the clocks of all the annelids including the well-known one of *Lumbricus terrestris*.

Chapter 6

CLAMS, OYSTERS, SNAILS AND
SEA HARES

THE PHYLUM MOLLUSCA, whose diversity is probably not rivaled by any other group of animals, and whose number of species— variously estimated between 65,000 and 100,000—stands inferior only to that of the Insecta, provides comparative anatomists, physiologists and embryologists with information and examples which illustrate important principles of their fields of inquiry. Ecologists and students of adaptations and evolution also find these animals, originally marine but having come to penetrate and occupy all habitats except the aerial one, invaluable and intriguing for their investigations. The gourmet, too, appreciates this vast assemblage, and fortunately for him, many of the edible species are sessile, sedentary or slow moving, although a few others, as the squid, are among the swiftest of marine animals.

Early in the present century, several biologists who were excited by the notion of living clocks approached molluscs with questions of biochronometry. They selected forms which inhabit the intertidal and littoral zones. These are the habitats of the majority of the members of the phylum. The choice of molluscs as animals with which to study temporal biological organization was and continues to be a wise one. In their natural environments, many of these animals live in time with the tides, the moon or the sun. In most parts of the world, especially near sea coasts and inland waters, representatives of the great phylum, Mollusca, are available in large numbers and can be kept under laboratory conditions for long periods of time.

During the last 20 years, more investigators of long-term physiological cycles have chosen clams, oysters, snails and sea hares as the subjects of their observations and experiments. We have found a variety of indicator processes, muscular activities, oxygen-consumption, patterns of spatial orientation and electro-physiological changes, which signal relationships with the solar, tidal and lunar cycles of our physical world.

Fascinating correlations between some of the molluscan rhythms and cycles of geophysical factors such as barometric pressure and natural radiations have been described. The phases and forms of cycles of several species have been changed after transport of the animals, by manipulations of light-dark periods, by variations in magnetic and electrostatic fields or by increases in levels of gamma radiation. Cyclic bioelectric activities in single neurons and in optic nerves are being elucidated by investigating neural units which fire at circadian frequencies.

Tidal and lunar cycles of marine bivalves and gastropods were among those reported in the literature of the early twentieth century. A snail, *Littorina rudis*, was found most active in the laboratory at 15-day or semilunar intervals (Bohn, 1904). In their natural habitats, these animals are covered by the sea only during the biweekly high, high tides. Under laboratory conditions, the average oxygen-consumption of two other species of snails and of a marine mussel was highest about the times of high tides and lowest around the times of low tides in the animals' native environments (Gompel, 1937).

In his report of 1918, Szymanski included activity records of common land snails, *Helix pomatia,* whose crawling was recorded under day-night changes in illumination and temperature, 48 hours at a time. The distribution of their periods of activity and rest was not regular through the day. However, the snails had the tendency to crawl maximally during early forenoon, late afternoon and evening. Now, in several European laboratories, details of rhythmic pacemaker activities of neurons are being analyzed in preparations of *Helix* nervous tissues (Ajrapetyan, 1973 and Neher, 1973). And, in Austria, a persistent annual cycle of oxygen-consumption has been recorded from relatives of *Helix, Arianta arbustorum* (Wieser, Fritz and Reichel, 1970). Whether the individuals had come from the highlands or the lowlands near Innsbruck, the gastropods used oxygen at the greatest rate during the summer months when maintained in the laboratory.

Rhythms of pumping in marine mussels, *Mytilus,* (Rao, 1954), of the opening and closing of the valves of marine clams or quahogs, *Mercenaria (Venus),* (Bennett, 1954) and of oysters, *Ostrea,* (Brown, 1954b) were among the first persistent cycles of bivalves to be reported in detail. Two species of mussels, *Mytilus californianus* and *Mytilus edulis,* some from subtidal and other individuals from intertidal habitats, propelled water through their mantle cavities at primary tidal frequencies. Peaks were recorded every 12.4 hours. The pattern continued for more than four weeks in animals which were maintained in the dark, in constant light or under natural day-night changes in illumination. Rao observed no evidence of a solar cycle.

The period of the tidal rhythm was temperature-compensated in the range of 9° to 20° C, and its phases could be set by exposing caged mussels to actual tidal environments. *Mytilus edulis,* collected off Cape Cod, were flown to California where they lived in the intertidal zone at Corona del Mar for a week. When they were moved to the laboratory, their maximal pumping occurred during times of the California high tides. Massachusetts animals which were studied in the laboratory without a sojourn in the Pacific ocean also functioned with tidal periodicity, but out of phase with the California tides. No shifting of the phases of the mussels' rhythms relative to tidal or lunar events was reported.

A "spontaneous" shift of phases of the tidal cycle of activity of oysters has been seen (Brown, 1954b). Those animals were collected from the New Haven area of Long Island Sound and were trucked to inland Evanston, Illinois, where they lived under constant laboratory conditions. As illustrated in Figure 30, the maxima of the oysters' rhythm were in synchrony with the times of high tides in the region of their collection, Long Island Sound, during the first semilunar (15-day) period of recording in Evanston (Fig. 30, A). However, during the second and third semilunar periods of study in Evanston (Fig. 30, B and C), the hours of greatest activity of the bivalves were those of maximal lunar gravitational attraction or the times of atmospheric high tides, in Illinois. Comparable lunar phases occur roughly one hour later in Illinois than they do in Connecticut. Brown found that the maxima of the oysters' 12.4-hour cycle moved spontaneously from times of New Haven high tides to times an hour later, and then persisted in that new relationship. The shift has been emphasized (Brown, 1960) as supportive evidence of the theory of regulation of persistent organismic cycles by exogenous factors.

But, the oysters' rhythms were average ones derived by the sliding method of Laplace, discussed in Chapter 3. Their amplitudes were low. The reality of these lunar cycles, the shifting of their phases and its significance have been met with great skepticism by many biologists. However, average 15-day solar cycles also calculated from the data of the oysters' activity showed changes in form from one semilunar period to the next, changes which could certainly reflect concurrent influences of lunar frequency, and thereby strengthen Brown's claims. The solar rhythms were characterized by three peaks of activity that moved, an hour or two back and forth relative to sun time, with succeeding halves of lunar-months.

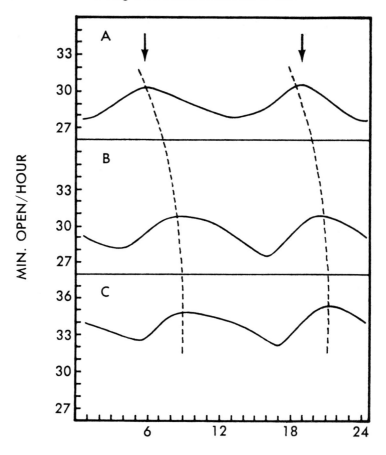

MIN. OPEN/HOUR

HOURS OF THE SOLAR DAY

Figure 30. Lunar rhythms of activity of oysters living under constant laboratory conditions in Evanston, Illinois. A. February 28 through March 14. The arrows indicate the times of high tides in the New Haven area native to these individuals. B. March 15 through March 29. C. March 30 through April 13. All dates were in 1954. Broken lines indicate the direction and extent of the "spontaneous" shifts (see text). From: Brown, F.A., Jr.: Persistent activity rhythms in the oyster. *Am J Physiol, 178*: 510-514, 1954b.

Should this particular translocation experiment not be repeated? Its results and their consequences are too important to our understanding of biochronometry for us to leave them in their ambiguous states. Barnwell (1973) is attempting even more detailed analyses of the rhythms of shell movements of oysters. He has found complexes of cyclic patterns of activity which could serve as the basis for more precise comparisons of the bivalves' rhythms after translocation.

Average cycles of the opening and closing of the valves of quahogs have also been described (Bennett, 1954). As were some of the mussels studied by Rao and the oysters observed by Brown, the animals whose activity I recorded were kept in the laboratory under constant conditions for as long as four weeks at a time. The movements of their valves were recorded automatically and continuously during month-long periods. The quahogs' solar cycle was simple in form (Fig. 31) with a clear

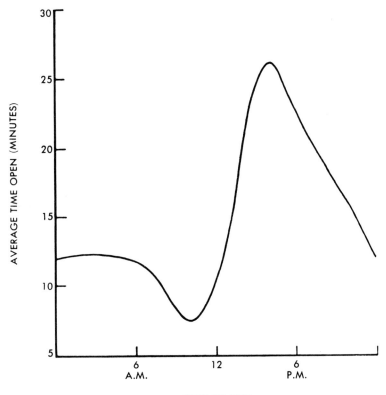

Figure 31. An average solar-day cycle of the opening of the valves of quahogs living under constant laboratory conditions in the spring of 1953.

maximum occurring between noon and late afternoon and a minimum apparent late in the morning. I suggested a positive and seemingly adaptive correlation between the phases of the clams' solar rhythm and phases of the diurnal cycle of vertical migrations of plankters, the organisms upon which the bivalves feed. Many more of these microscopic creatures are found near the bottom of the sea, where quahogs live, from morning until evening than during the night.

In addition to the solar cycles, the activity data of my study traced out an obvious long-term cycle (Fig. 32). Phases of this cycle correlated with lunar events. I postulated that the long-term cycle was one of lunar-monthly frequency, 29.75 days, whose major and minor peaks of activity alternated at approximately 15-day intervals, and were synchronized with the days of lunar zenith and nadir in the quahogs' native environment. The cycles of the animals whose ecological histories were known supported those ideas.

Statistical analyses of the activity of quahogs observed later hinted at a possible 27-day cycle (Brown, Bennett, Webb and Ralph, 1956). Similar indications were seen for oysters whose shell movements were recorded at the same time and in the same laboratory as those of the quahogs. Rhythms of that period have

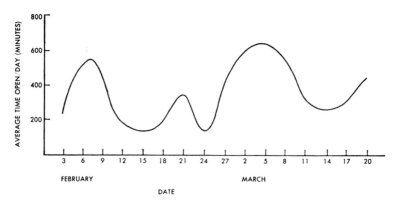

Figure 32. The long-term or lunar-monthly cycle of the opening of the valves of quahogs living under constant laboratory conditions in the spring of 1953. From: Bennett, M.F.: The rhythmic activity of the quahog, *Venus mercenaria*, and its modification by light. *Biol Bull, 107*: 174-191, 1954.

been suggested by others who also note that the frequency of the cycle of the rotation of the sun on its own axis is 27 days. Simpson (1954) has documented and explained 27-day cycles in the levels of some components of cosmic radiation. Does the effector of geophysical rhythms of 27 days also stimulate living systems and allow them to generate physiological cycles of that same period? Here is another question whose answer may well demand concerted and cooperative efforts of biologists and geophysicists.

Average 15-day, tidal or lunar-day rhythms, comparable to those reported for the oysters from Long Island Sound, were also established for quahogs (Fig. 33). The two minima of the clams' activity rhythm, periods when their shells tended to be closed, fell within an hour of the times of low tides in the areas of the animals' collection. Quahogs living in the upper reaches of the intertidal zone are often uncovered at low tides as the waters

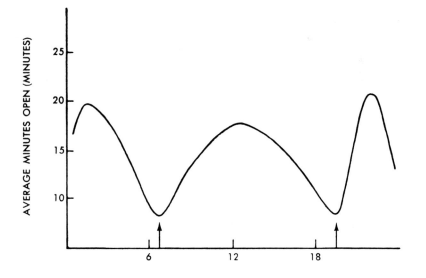

HOURS OF THE SOLAR DAY

Figure 33. An average 15-day tidal cycle of the opening of the valves of quahogs living under constant laboratory conditions in the spring of 1953. The arrows indicate the calculated times of low tides in the animals' original habitat.

recede. During those times, bivalves typically clamp their shells together. The periods of maximal opening included the hours of high tides in the bay from which the clams had been dredged. Bivalves usually open their shells, and pump water through their mantle cavities when they are well covered by the sea. Therefore, a tidal cycle characterized by maximal opening during high tides and minimal opening during low tides would be adaptive to marine clams such as the quahogs.

The phases of the three cycles which I described, solar, tidal and monthly, were shifted relative to real time and to comparable phases of the rhythms of control animals, which lived under constant low illumination, by exposing quahogs to alternating periods of darkness (8:00 AM to 8:00 PM) and light (8:00 PM to 8:00 AM) for five consecutive days. The L/D regimen did not cause simple reversals of the cycles, but the shifts which did occur showed that such perturbations might entrain the rhythms of this particular mollusc.

That conclusion is supported by very recent findings of Barnwell and his associate who are analyzing shell movements, shell growth and photoperiods in quahogs as well as in oysters (Barnwell, 1973). However, changes in photoperiod have been claimed to have no effect on the banding or pattern of growth lines in some other bivalves, both mussels and scallops (Dodd, 1969). Dodd believes that this banding is an indicator of circadian rhythmicity, whereas Evans (1972) has found that the pattern of growth lines of a basket cockle, *Clinocardium nuttalli*, reflects tidal rhythmicity.

The detailed study of shell growth in lamellibranchs not only provides students of biochronometry with valuable information, but also investigations of the phenomenon benefit students of geochronometry. Were the solar and lunar periods of our geological past the same as those of the present? Some geologists think not. They believe that a year consisted of a greater number of shorter days than it does now, and hope to prove their ideas by checking growth lines of fossil molluscs whose patterns might reflect the frequencies of ancient geophysical cycles.

The solar and lunar activity rhythms of present-day quahogs

and oysters show interesting relationships with modern geophysical cycles. Many of these correlations are the same types which have been demonstrated for cycles of oxygen-consumption in fiddler crabs (Chapter 3) and in salamanders (Chapter 7).

Of course, such relationships do not prove that geophysical factors such as barometric pressure and cosmic radiation, whose levels correlate with those of the animals' activities, induce the organisms' rhythms. It has been argued, however, that an unknown external factor or a combination of factors whose fluctuations parallel or even cause oscillations of barometric pressure or components of cosmic radiation may be causally related to the organismic cycles. Whatever the truth may prove to be, these correlations indicate the possibility of geophysical influence in the genesis and modulation of some biological timing mechanisms. Further, the relationships suggest the probability that living systems are able to use one or a group of many different geophysical forces as Zeitgebers or entraining agents. Changes in light intensity, temperature or sound are undoubtedly not the only variations which set components of organisms' clocks in time with events of our normal, fluctuating physical environments.

The forms of the cycles of shell movements of oysters and quahogs were compared with concurrent changes in barometric pressure (Brown, Bennett, Webb and Ralph, 1956). For both species, there was a general tendency for the animals to open their valves for longer times as the barometric pressure fell, and conversely, to keep them closed as the pressure rose. The variations in another geophysical factor, the nucleonic component of cosmic radiation, also correlated with differences in the shell activities of the molluscs. As illustrated in Figure 34, the relationship was a negative one, for the major parts of the quahogs' cycles were inverted relative to those of cosmic radiation, a situation similar to that described in Chapter 3 for fiddler crabs' metabolism and the nucleonic component. In addition, as with the fiddler crabs, the form of the solar rhythm of the quahogs inverted "spontaneously" between the summers of 1954 and 1955. The pattern of the daily cycle of the nucleonic component of

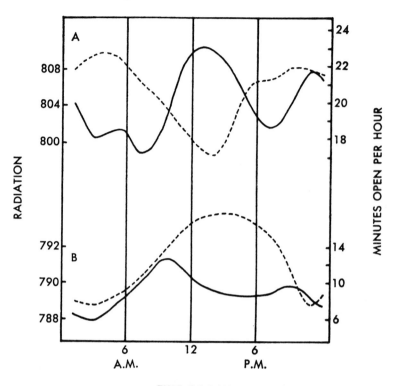

Figure 34.　A. The 3-hour moving mean of the average solar cycle of the nucleonic component of cosmic radiation for July 7 — August 4, 1954 (broken line) and a 3-hour moving mean of the average daily cycle of opening of the valves of quahogs for the same period (solid line). B. The same as A., but for July 6 — August 3, 1955. From: Brown, F.A., Jr., Webb, H.M. and Bennett, M.F.: Comparisons of some fluctuations in cosmic radiation and in organismic activity during 1954, 1955 and 1956. *Am J Physiol, 195:* 237-243, 1958.

radiation did also. Consequently, the relationship between the two cycles—organismic and geophysical—was again seen to be grossly the same during two years, 1954 and 1955. In our 1958 publication (Brown, Webb and Bennett, 1958, p. 242), we pointed out that it seemed, ". . . probable that the organism is dealing with some physical factor whose intensity correlates with that of cosmic radiation as a consequence either of a cause-effect relationship between them or of their fluctuations being parallely modified by a third factor." That factor or group of factors still eludes us, and the mechanism of the living clock of marine bivalves continues to be unknown.

Investigators of the Hungarian Biological Research Institute on Lake Balaton have published information about timing phenomena in fresh water bivalves, *Anodonta cygnea* (Salánki and Vero, 1969). These lamellibranchs have a diurnal rhythm of activity, mediated in part by a chemical agent of the central nervous system. The agent may be serotonin, 5-hydroxytryptamine (5HT), for increases in the concentration of 5HT or its precursors increase muscular activities of the mussels, while inhibitors of 5HT metabolism decrease the animals' activity.

The cyclic activities of a marine gastropod, the common mud snail, *Nassarius obsoletus,* are other indicators of the clock of the Mollusca. This snail, a member of a family represented throughout the world, is one of the most abundant species of the Atlantic coast. It is predaceous, feeding while moving over submerged mud flats in its natural surroundings. Under constant conditions of the laboratory, mud snails crawled maximally at times of high tides on their native beaches, and were least active during low tides (Stephens, Sandeen and Webb, 1953). Not only does the amount of crawling vary rhythmically in *Nassarius*, but also the rates at which the animals moved differed from hour to hour during the daytime (Webb, Brown and Brett, 1959).

These snails were also tested under constant conditions. Their greatest rates of locomotion were recorded around 10:00 AM and from 2:00 PM until 4:00 PM. The snails were slowest near 8:00 AM, noon and 6:00 PM. Imposed electrostatic fields reduced the rates of crawling during the middle of the morning, but did not have obvious effects on the indicator process in the afternoon. Related studies with a flatworm, *Dugesia dorotocephala,* provided evidence that its orientational responses to electrostatic forces differed in the morning and afternoon, too. Are electrostatic fluctuations in the earth's atmosphere regulators of animals' clocks? Because of organisms' varying sensitivity to them, do they, like changes in light and temperature have unlike effects on living systems during various periods of the day?

Some atmospheric electrostatic changes occur on universal

time, and therefore, exert their influence everywhere on the face of our earth at virtually the same second. Did those changes keep our Massachusetts crabs on Atlantic Coast time after their flight to California (Chapter 2)? Were Renner's bees, trained in France and tested in New York, cued by electrical variations which could have been measured synchronously in Paris and in the American Museum of Natural History (Chapter 4)? If so, how can we explain the Connecticut oysters' shifting the phases of their primary lunar cycle after two weeks in Illinois? Were they responding to geophysical factors whose fluctuations do not occur on universal time? Our knowledge of even basic interactions of living systems and subtle geophysical factors is severely limited. Study of such interactions may become "the" field of investigation for the new generations of physiologists.

In addition to contributing information about rates of crawling and a subtle geophysical factor—electrostatic flux—mud snails have taught us about persistent cycles of oxygen-consumption (Brown, Webb and Brett, 1959). Similar cycles have been described for two other species of marine gastropods, the periwinkle, *Littorina littorea,* and the oyster drill, *Urosalpinx cinereus* (Sandeen, Stephens and Brown, 1954). Under conditions of constant light and temperature, all these intertidal gastropods metabolized at solar and lunar frequencies. The forms of their circadian cycles were species-specific as were the patterns of their tidal rhythms. The data recorded for the mud snails are especially valuable, for the animals were hermetically sealed away from atmospheric pressure changes. Consequently, their constant laboratory conditions included constant barometric pressure. Around 6:00 AM and 6:00 PM, the levels of oxygen-consumption of *Nassarius* correlated with weather related, and hence, unpredictable, fluctuations in barometric pressures at the same hours—even though the snails could not have experienced the changes in atmospheric pressure.

This twice daily correlation between oxygen-consumption and barometric pressure was first recognized by Brown and his associates in pieces of potato tubers, maintained in respirometers under constant conditions which also included constant pressure (Brown, 1958). Which geophysical factor affects the weather, snails and potatoes in such a manner that their reactions to it are manifest around 6:00 in the morning and in the evening?

Another relationship between barometric pressure and levels of metabolism that has been found for both mud snails and potatoes is this: the average oxygen-consumption of the organism on day n correlated with the average barometric pressure on day n plus 2. If the organisms use fluctuations in external factors to help themselves tell time, the variations which today regulate the living systems must influence our physical surroundings in such a manner that barometric pressures reflect that effect not today, but 48 hours from now, or two days hence. Can plants and animals forecast the weather for us? The types of statistical correlations cited here are some that supporters of the theory of exogenous modulation of biological clocks often offer as proof of their ideas. However, many other biologists believe such numerical relationships to be merely fortuitous.

Spatial as well as temporal patterns of orientation of the same mud snail, *Nassarius obsoletus,* have been claimed to be influenced by changes in geophysical factors, especially magnetic force. These claims are backed by literally hundreds of thousands of single observations which have been painstakingly analyzed by Brown, and have been incorporated as part of his theory of the mechanics of biological timing (Brown, 1970). If Brown's conclusions about the snails' orientation prove to be correct, the idea that cyclic fluctuations in external environmental factors are necessary for the genesis and maintenance of persistent biological rhythmicity—originally proposed by Stoppel—will have gained tremendous, perhaps irrefutable support.

The study of orientation of *Nassarius* started in Brown's Woods Hole Laboratory in the summer of 1959. We compared the paths of mud snails in the earth's magnetic field with those of snails exposed to experimental fields. The experimental fields, which were nine to ten times the intensity of that of the earth, were placed parallel to or at right angles to the geomagnetic field. The animals were observed hourly from 5:00 AM to 9:00 PM. Thirty-three thousand, eight hundred and forty snail paths were scored as the animals crawled from a narrow canal, the exit of which pointed toward the true magnetic south, out into an arena with a white substratum and with diffuse, and therefore, non-biased illumination (Fig. 35). As illustrated in Figure 35, that arena was divided into sectors, each of 22.5 degrees. The exit from

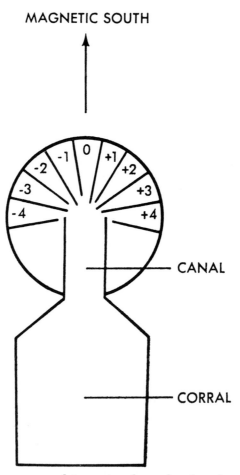

Figure 35. The arena used in scoring the paths of mud snails. The snails are placed in the corral, and then are observed as they crawl from the canal onto the white background marked in 22.5° sectors. From: Brown, F.A., Jr.: Response to pervasive geophysical factors and the biological clock problem. *Cold Spring Harbor Symp Quant Biol,* 25: 57-71, 1960. Copyright 1961 by Cold Spring Harbor Laboratory.

the snails' canal was lined up with the zero sector, so that the plus sectors lay to the right and the minus sectors to the left of the exit. Therefore, the paths of the animals could be scored in terms of the direction of turning and the amount of turning from magnetic south. The bar magnets used to modify the earth's field were on moveable platforms under the orientation arena, and were easily aligned as indicated by protocol.

The snails moving in the earth's field varied their paths cyclically (Fig. 36). In terms of solar time, they moved straight ahead, *i.e.*, toward the south, early in the morning. From then until midday, they turned increasingly to the left or eastward (− in the figure). During the afternoon and evening, the degree of left-turning decreased. In terms of lunar time, the minimal degree of left-turning was recorded midway between lunar nadir and zenith, and the peaks of left-turning were observed at lunar nadir (Fig. 36).

The snails crawling in the two different experimental magnetic fields also changed their paths rhythmically. The temporal patterns were much the same as those of *Nassarius* tested in the earth's field, but the experimental animals always turned to the left of the controls. Further analyses (Brown, Bennett and Webb, 1960) suggested that the snails were able to differentiate between the directions of the two experimental fields, the one parallel to and the one at right angles to the earth's field, and that their capacity to differentiate varied with solar and lunar time. Around solar noon and near the time of lunar nadir, that capacity was maximal.

We emphasized that the possession of such living compasses, hinted at in our analyses, along with the living clocks, indicated by the several sets of hands of the *Nassarius* clock, could ". . . constitute a potential means for organismic navigation in the absence of more obvious cues" (Brown, Bennett and Webb, 1960, p. 73). If mud snails actually possess this type of navigational system, do they use it? One would guess not. Properly speaking, *Nassarius* do not navigate. They crawl about in a crisscross pattern in all directions through very limited areas of their mud flats. Much of the time, their movement appears to be undirected. When

LUNAR TIMES

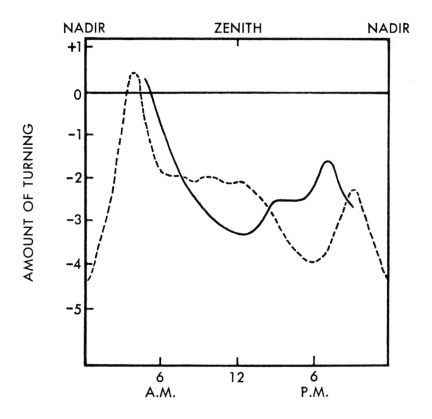

Figure 36. An average solar-day (solid line) and an average lunar-day (broken line) curve showing the amount and direction of turning of mud snails from a southbound path. + and − correspond with the + and − sectors of the orientation arena (Fig. 35). From: Brown, F.A., Jr.: Response to pervasive geophysical factors and the biological clock problem. *Cold Spring Harbor Symp Quant Biol*, 25: 57-71, 1960. Copyright 1961 by Cold Spring Harbor Laboratory.

food—living or dead and decaying animal material—is nearby, individuals or aggregations of snails move directly and rapidly toward it. The response is assumed to be a type of chemical orientation.

How fine it would be if clear and obvious reactions of truly navigating species to magnetic fields and changes in them were to be found. Scores of ornithologists have postulated a role of geomagnetism in the navigation of migrating birds (Presman, 1970), but as noted earlier, in only a few cases has positive evidence of such a reaction been secured (Keeton, 1972 and Southern, 1970 and 1972). Modification of the dances of honey bees by manipulation of the magnetic fields to which the insects were exposed (Chapter 4) proves that worker bees can react to that geophysical force. However, to date, no one has demonstrated that *Apis* uses the capacity in directing its flight between the hive and sources of nectar and pollen. Noted in Chapter 5 were examples of the effects of geomagnetism on the temporal organization of animals so different as earthworms and man. Is it not probable that the mud snail's ability to differentiate among characteristics of magnetic fields is an essential link in its timing rather than in its orientational mechanism?

Brown and his students have continued the study of relationships of geomagnetism to both animal orientation and animal clocks. *Nassarius* and flatworms, *Dugesia*, have yielded much of the information with which these investigators have worked. More evidence that mud snails can differentiate among compass or geographical directions was adduced, and seasonal variations in the average paths of the animals were noted. The scoring of 51,040 snail paths showed that a reversed magnetic field of a strength very close to that of the earth, 0.17 gauss in Massachusetts where the study was made, evoked greater responses than did reversed fields of much greater forces. Whatever the magnetoreceptor mechanism of *Nassarius* is, its sensitivity seems to be adjusted precisely to forces very close to those of the horizontal vector of geomagnetism, *i.e.*, to forces which actually exist in the environment of animals on this earth.

Studies with *Dugesia* (Brown and Park, 1965a and 1965b) strengthened the conclusions based on the observations of the mud snails, and the flatworms and the same species of gastropods have also proved to us that temporal and spatial aspects of animal orientation can be affected by weak gamma radiation (Brown, 1969).

Syntheses of all these investigations resulted in Brown's

theory of the Space-Time Continuum in the regulation of persistent cycles of daily, tidal, monthly and annual frequencies. He has emphasized (Brown, 1969, p. 294) that:

> The spatially and temporally related subtle geophysical variations in the environment of the organisms form a continuum (Fig. 37). At any given moment in time the subtle geophysical field varies with geographical direction. For any fixed geographic point there is a continuous variation in the field with time. Essentially the same earth's electromagnetic parameters which provide subtle geographic directional information can, therefore, theoretically provide temporal information within the frame work coordinates of the natural geophysical cycles.

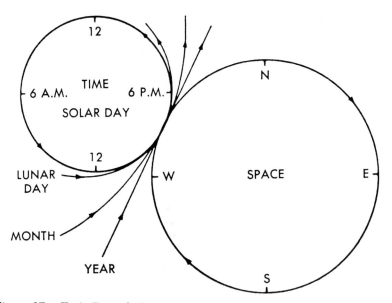

Figure 37.   F. A. Brown's Space-Time Continuum. "At any instant in time there is a variation in space, here represented by the 360° geographical cycle. Each point in space is varying with time including solar-day, lunar-day, monthly and annual periodic components." From: Brown, F.A., Jr.: Hypothesis of environmental timing of the clock. In Palmer, J.D. (Ed.): *The Biological Clock. Two Views.* 1970, pp. 13-59.

There are probably no major objections to these statements. But, is Brown's hypothesis actually supported by the evidence at hand? The orientational differences which have been reported for snails, flatworms and some protistans are very slight, quantitatively speaking; they have been based on averages which represent extremely wide ranges; they have not been confirmed by other workers. Further, even if such mechanisms are actually used by *Dugesia* and *Nassarius,* do all organisms which show persistent rhythms under constant laboratory conditions depend upon the factors which constitute the Space-Time Continuum for the generation or even the regulation of the frequencies of their cycles? Or, can other organisms run the works of their living clocks without cues from their rhythmic external worlds? Basic questions are faced once again: Are living clocks completely endogenous? Are they completely exogenous? Have both types and mixtures of them evolved on this earth?

Fascinating information about the electrical activity of some living clocks has come to us from sea hares, *Aplysia,* which are shelless relatives of marine snails. The sea hares have a circadian rhythm of locomotion, and their contributions to biochronometry have been received with excitement. Students of pacemaker phenomena and neurophysiologists, in general, have shared that excitement, for they, too, value the elegantly precise data being gleaned from cells of these large, seaweed grazing animals. Particular giant cells in clusters of nerve cells or ganglia of *Aplysia* can be identified specifically in all specimens. For example, the cells of the parieto-visceral ganglion (PVG) are known well. The output of individual cells has been recorded, and the records are available for study and comparison (Strumwasser, 1965 and 1967).

The records are repeatable, assuring us that comparing electrophysiological data from the same cell of different animals at various times and under a variety of experimental conditions is valid. Cell 3 of the PVG has come to be referred to as the parabolic burster (PB). It emits "spontaneous" bursts of roughly 12 spikes which are separated from one another by silent periods (Fig. 38). When the successive spike intervals of a burst are plotted against time, the resulting curve is parabolic. Hence, the name parabolic burster was assigned to cell 3 of the PVG.

For observations of the temporal performance of PB cells

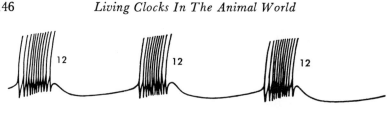

—————————— 20 SECS.

Figure 38. A diagram of the typical pattern of spikes produced by a Cell-3 or Parabolic Burster (PB) of an *Aplysia* parieto-visceral ganglion. From: Strumwasser, F.: The demonstration and manipulation of a circadian rhythm in a single neuron. In Aschoff, J. (Ed.): *Circadian Clocks.* 1965, pp. 442-462.

and for experiments on their circadian performance, PV ganglia are removed from sea hares, and are kept in glass containers at constant temperature while being perfused with sea water or a test solution. Under these conditions, the possible influences of receptors, hormones, extraganglionic neurosecretions or any other substance of the animals' normal body fluids are eliminated. The tips of microelectrodes are placed inside the PB cells, and consequently, intracellular electrical changes can be recorded. Most preparations of the cells are good for as long as 48 hours of recording. While living in aquaria, intact sea hares, the potential donors of the groups of cells, can be conditioned to appropriate light-dark regimens for a selected number of days before their cells are isolated and studied *in vitro.*

After *Aplysia* have been exposed to 12 hours of light followed by 12 hours of darkness, the greatest rate of production of spikes by PB cells is usually seen within a short time of "projected" dawn on the first day after the isolation of the cells (Fig. 39). "Projected" dawn is the solar time at which the intact sea hares experience a dark to light change. Approximately 24 hours later on the second day of isolation, a second period of intense activity is recorded. The circadian frequency of this cycle of electrical activity is also obvious in records of PB neurons from *Aplysia* which live under constant light for a week before their ganglia are removed and prepared for investigation.

Another giant cell which functions with circadian frequency is cell 15 of the abdominal ganglia of sea hares (Lickey, 1969). Very obvious seasonal differences in this cell type's circadian cycles have also been described. From November through March, peaks of electrical activity were recorded at dawn (lights-on) and at dusk (lights-off). During that part of the year, the cycle could be entrained to 21-, 24-, or 27-hour periods by exposing intact animals to L/D 10.5/10.5, 12/12 or 13.5/13.5 before their ganglia were removed. In September and October and again in April through June, the greatest electrical activity occurred at solar midday and midnight, and entrainment to the 27-hour day was not possible.

Molluscs of the genus *Aplysia* have still a third set of hands

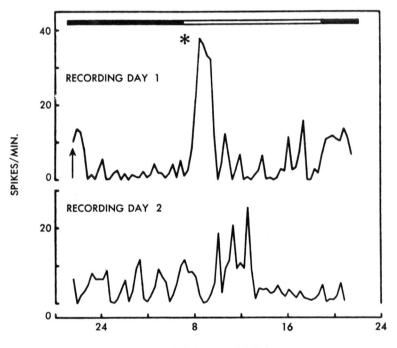

CLOCK TIME, HOURS

Figure 39. The output of a PB cell in terms of clock time. The black and white bands above the record of Day 1 indicate the periods during which the sea hare donor had experienced dark and light for the 9 days preceding the days of recording. The arrow indicates the time at which the electrodes were placed in the cell. * marks the time of "projected" dawn. From: Stumwasser, F.: The demonstration and manipulation of a circadian rhythm in a single neuron. In Aschoff, J. (Ed.): *Circadian Clocks.* 1965, pp. 442-462.

of their biological clocks. The electrical changes recorded from the stump of an optic nerve attached to an eye isolated from a sea hare vary at circadian frequency (Jacklet, 1969). This pattern of activity has been found in preparations maintained in constant darkness, in constant dim light, at constant temperatures ranging from 15° to 25° C, in sea water or in a sterile culture medium. In sea water, eye-nerve preparations from donors which had lived under constant low illumination had free-running periods of less than 24 hours, while those from sea hares which had been exposed to L/D 12/12 had cycles very close to 24 hours in length. However, preparations kept in the sterile culture medium showed free-running periods of 27 hours even though their donors, too, had lived under L/D 12/12. The latter preparations do not deteriorate so fast as those maintained in sea water.

This indicator of the clock of *Aplysia* would seem to be especially valuable for investigations focused on the effects of metabolic stress, the state of nutrition or the influence of toxins on timing mechanisms. It is also elegant for studies of phase-shifting and entrainment by light-dark changes both *in vivo* and *in vitro* (Jacklet, 1969 and Eskin, 1971). *In vivo*, the cycle of one eye is phase-shifted, while that of the second eye, which is capped during light perturbations, is not. Consequently, experimental and control preparations can be secured from the same individual, a situation not always possible for experimentalists.

Using these isolated eyes of *Aplysia*, Jacklet and Geronimo (1971) are analyzing the interaction of the many neurons of the photoreceptors in the production of circadian rhythms. A population effect has been discovered. As the number of neurons of a preparation is reduced below 20 per cent by cutting away portions of the retina, shorter circadian periods are recorded. Eyes with as few as 20 per cent of their neurons have frequencies averaging 25 hours; those with only 10 per cent of these cells have periods averaging 21 hours; those with only 5 per cent have periods of 7.5 hours. Whole eyes expressed cycles of 26 to 28 hours in length. Complete explanations for the interactions which generate the circadian frequencies have not been worked out, as yet. Since the clock of *Aplysia* has several indicators—the spontaneous electrical activity of giant neurons, bioelectrical changes in optic nerves and levels of locomotory activity, the sea hares are also potentially effective systems with which to analyze interactions among the various sets of hands of a living clock and those between the hands and the clock itself. These relationships await analysis.

A very fundamental question of biochronometry which has been put to *Aplysia* is: what is the nature of the temporal control of a single cell, in this case the parabolic burster neuron? A theoretical answer has come from Strumwasser's work (Strumwasser, 1965). His hypothesis implicates many structures and functions of the cell—its membrane, enzymes and their actions, DNA and RNA and their activities in protein synthesis. Strumwasser stressed his belief that the parabolic burster's electrical activity is endogenous to the cell, or is certainly not induced by synaptic input. His experimental results support the conclusion that the PB cell does produce "on clock time" a depolarizing substance, the DS. This material, a polypeptide or a protein, depolarizes the inner surface of the membrane of the cell in which it is produced, and thereby evokes the electrical changes which are reflected by the spikes whose frequencies vary with solar periodicity. The DS, or the enzymes necessary for its production, are synthesized in the PB cell under the direction of specific messenger RNA. Following the now familiar story of molecular biology, the messenger RNA is synthesized under the direction of specific DNA or genetic material. Therefore, the basis of the circadian changes in this giant neuron resides in the nuclear control of protein synthesis.

What types of evidence support these ideas of Strumwasser? Some comes from results of investigations of the effects of brief increases in temperature on PB cells, *in vitro*. Electrical changes in the neurons were recorded before, during and after the experimental treatment. The "heat pulses" caused advances in the times of the PBs' peaks of activity, *i.e.*, the peaks came earlier than was usual. The advances were explained as follows: the higher temperature accelerated the release of messenger RNA which had been synthesized and was stored in the nucleus. Since the RNA was available earlier than usual, DS was synthesized in the cell before the expected time in the circadian period. Some of the effects of actinomycin D, an antibiotic, on the rhythm of electrical activity mimiced those of the heat pulses. The general effects of actinomycin D are on protein synthesis, and its specific effects on the PB cells' rhythms were related to the drug's disruption of the timing of the production of DS, postulated to be a protein or at least a polypeptide. These preliminary studies cause one to be optimistic that *Aplysia* may supply us with fundamental information about the biochemistry of circadian systems. Until recently, much of that knowledge was contributed by unicellular organisms, *Paramecium, Tetrahymena,* and *Gonyaulax,* and as was pointed out in Chapter 1, the facts of the biochemistry of living clocks are few.

Meager though they are, these facts consistently emphasize the involvement of proteins, especially enzymes, in the timing mechanisms of cells. In Hasting's laboratory at Harvard (Hastings, 1970), the activity of luciferase, an enzyme necessary for the emission of light in *Gonyaulax,* has been found to be greater at night than during the day. Parallelly, under constant laboratory conditions, this bioluminescent, marine organism flashes more often and more brightly during the nighttime than in the daytime. The concentration of a second enzyme, ribulose diphosphate carboxylase, also varies in *Gonyaulax* during the circadian period. This enzyme catalyzes photosynthetic reactions whose levels fluctuate with solar time. Photosynthesis reaches its maximal rate during the middle of the day; the concentration of the specific carboxylase is greatest at the same time. How are these changes in concentrations caused? Temporal differences in the syntheses of specific enzymes constitute one possible explanation.

Although it has loopholes, and cannot explain all details of biochronometry, the chronon hypothesis of Ehret and his associates at the Argonne National Laboratory (Ehret and Trucco, 1967) could explain rhythmic variations in the syntheses of enzymes or any other cellular proteins. Ehret's model assigns to subunits of chromosomes or particular long segments of DNA the role of cellular timepiece. Each of the subunits or segments is called a chronon, and consists of a substantial number of genes in precise linear order (Fig. 40). One end of a chronon is occupied by the initiator gene (gene 1 of Fig. 40), and the opposite end by the terminator gene (gene 6 of Fig. 40). The genes of any one chronon are transcribed in their linear order. The serial transcription of one chronon takes approximately 24 hours. Therefore, during a solar-day, the concentrations of specific messenger RNAs (mRNA 1 through mRNA 6 of Fig. 40) vary, and so do those of the proteins (Proteins 1 through 6 in Fig. 40) whose syntheses are directed by the specific molecules of RNA.

The messenger of the terminator gene somehow insures that

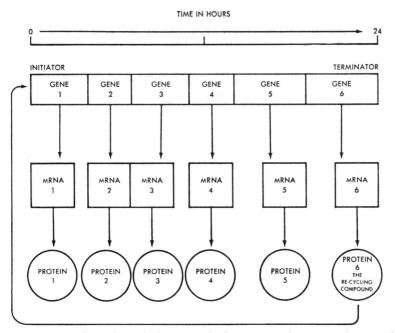

Figure 40. An hypothetical chronon which consists of genes 1 (initiator) through 6 (terminator). The genes are normally transcribed in linear order, so messenger RNA_1 through messenger RNA_6 are present in maximal concentrations in the cell one after another in time. Consequently, sequential syntheses of proteins 1 through 6 are directed. The transcription of this hypothetical chronon takes 24 hours, and is reactivated by the recycling compound produced under the influence of messenger RNA_6, the messenger of gene 6, the terminator gene.

the transcription of the linearly arranged genes of its chronon is reinitiated. The $mRNA_6$ may stimulate the production of a re-cycling compound, a protein, which activates the initiator segment of the chronon. Another circadian cycle is thereby started. Theoretically, the cycling would continue through the life of rhythmic organisms. Chronons longer than those whose transcription occupies about 24 hours could pace cycles of monthly or annual frequencies. Conversely, shorter chronons could account for 12.4-hour or primary tidal cycles of biological processes.

The Argonne investigators, using the unicells, *Paramecium* and *Tetrahymena*, have started the difficult and time-consuming task of proving whether or not specific messenger RNA molecules are present at particular times of the solar-day (Barnett, Ehret and Wille, 1971). If the chronon hypothesis explains timing in these protozoans, will it be applicable to all organisms? Will it be applicable to all the cells of a single multicellular organism, such as the sea hare, *Aplysia*, whose giant neurons have taught us that genes, their messengers and the molecules whose production they direct, must be reckoned with as we go on with our analyses and syntheses of the many diverse aspects of biochronometry?

Chapter 7

FROGS, TOADS AND SALAMANDERS

Some amphibians ". . . have literally used their heads to find microenvironments where conditions are favorable for survival and reproduction" (Bentley, 1966, p. 623). Do they, perhaps, also use their living clocks and calendars to adapt to their surroundings whether they be of a desert, a rain forest, a woodland pool or the tundra? All the animals of the small class, Amphibia, cold-blooded forms which are transitional between the aquatic and the terrestrial vertebrates, face the rigors of their environments—dry, damp or wet, cold, warm or hot—with naked moist skins. Therefore, one is often quite honestly amazed to learn of the wide distribution of the group, notably that of the frogs and toads. While most salamanders live in damp terrestrial habitats or in bodies of fresh water of the northern hemisphere, their tailless relatives, frogs and toads, are found throughout much of the world. Only the seas, the antarctic and the majority of oceanic islands are devoid of amphibians.

Especially in the drier, warmer parts of their ranges, many of these animals actually use their heads to burrow into the ground where they spend most of the day. They emerge to feed, to move about and to reproduce during the cooler, damper parts of the circadian period. However, nocturnalism is typical not only of the desert inhabiting Amphibia. Many amphibians which live in woods, in meadows or in ponds are also more active during the night than during the day. One suspects that biological clocks assure all amphibians of being in the right place at the right time of the solar-day.

The Amphibia of the temperate zones must also be in the right place at the right time of the year. They spend their winters in the mud at the bottoms of ponds or streams or in very damp terrestrial niches. During the spring, they come into reproductive condition and spawn in temporary or permanent bodies of fresh water. Their summers are spent feeding and preparing, physiologically, for the long period of dormancy of the colder months of the year. Countless field observations have confirmed both daily and annual variations in the lives of frogs, toads and salamanders.

Hints of the persistence of such variations under laboratory conditions are numerous. Investigators have been vexed and frustrated by the many physiological differences between "summer" and "winter" frogs. The results of studies conducted during these two periods of the year under the same experimental conditions with the same species and identical protocols are often unbelievably different. Results recorded at various times of the day also vary. In our laboratory, it was found that the hearts of common frogs, *Rana pipiens,* react differently to identical doses of a drug at midday and in the evening. What specific information do we have about persistent rhythms in amphibians? We have little.

One may be surprised to learn of the dearth of information about timing mechanisms in this group of vertebrates, because our knowledge of many aspects of the biology of the amphibians is so detailed. Biologists start to amass information about frogs, toads and salamanders early in their careers. Most first year college students know something of the developmental history of amphibians and something of the classic work of the frog and salamander embryologists.

Beginning zoologists often feed thyroid glands or thyroid extracts to tadpoles, watch them metamorphose prematurely and read of the early endocrinologists who chose frogs, toads or salamanders as their experimental animals. The anatomy and the histology of frogs are taught world-wide. The twitches of their gastrocnemius muscles are recorded millions of times per year in student laboratories. Indeed, amphibians have been and continue to be "martyrs to science" (Noble, 1931).

One amphibian which is currently being a martyr to developmental biology is the African clawed toad, *Xenopus laevis*. It also indicates the possession of a biological clock by changing color with circadian frequency under laboratory conditions of constant darkness and constant temperature. Hogben and Slome (1931) described that rhythm in a long paper concerning the details of coloration in *Xenopus*. These investigators described the periodic change in pigmentation simply as the tendency of melanin to be more concentrated at midnight than at noon. Whether their skins were light or dark when the *Xenopus* were placed under constant conditions, the animals' pigmentary variations were much the same. They were darker at noon than at midnight (Fig. 41).

Similar temporal variations were reported for frogs in a lengthy monograph by Minkiewicz (1933). He described the nocturnal phase of that cycle as the retraction of pigment. The rhythm of color change was said to be intrinsic, autonomous and to be influenced by a complex of factors including internal ones. In addition, Minkiewicz discussed the establishment of the cycle of color change. He found that light intensity, the duration of the light periods and the numbers of "schooling" (light-dark) periods to which the animals were exposed had effects on the

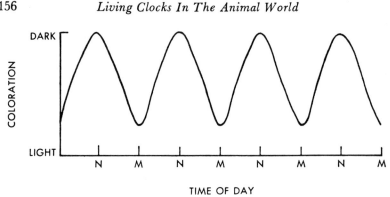

Figure 41. Daily variations in the coloration of *Xenopus* living under constant laboratory conditions.

persistence of the cycle. Has no one gone on with studies of rhythms of color change in the Amphibia? One might well postulate that many characteristics of those cycles, *e.g.*, hormonal mediation, phase-shifting and varying sensitivities to light and temperature perturbations, are similar to those of the chromatophore rhythms of fiddler crabs (Chapter 2). Even though the cycles of color change of amphibians are not so precise as those of the crustaceans, more detailed analyses of chromatophore rhythms of frogs and toads might provide valuable information about the clocks of these fascinating transitional vertebrates.

As do fiddler crabs, earthworms and molluscs, some amphibians indicate their possession of living clocks by periodically varying the levels of their locomotory activity under constant laboratory conditions. However, the first published report of an attempt to find such a cycle was negative (Szymanski, 1914). From November through January, fire salamanders *(Salamandra)* lived in chambers of Szymanski's recording instruments. During all that time, the animals moved so little or the actographs, which were certainly less sensitive than comparable instruments used now, picked up so few movements that the Viennese investigator concluded that the salamanders were completely quiescent. Perhaps they actually were, and perhaps that period of inactivity was a phase of an annual cycle. If those salamanders were observed under laboratory conditions during the spring and summer, would they be more active than they were found to be in the fall and winter? If they were active, would the degree of their activity vary with circadian frequency?

Positive evidence of persistent rhythmic activity in some other salamanders has been published. Larval axolotls living in the chambers of actographs showed a pronounced 24-hour cycle (Kalmus, 1940). Whether they were exposed to constant darkness or to alternating periods of light (by day) and dark (by night), the young animals were more active during the night than during the day. Kalmus also found that the phases of their cycles could be reversed by exposing the salamanders to bright light by night and to darkness by day. After that treatment, the larvae were less active during the nighttime than during the day. Thyroidectomized larvae behaved as did the normal ones, but hypophysectomized young, although rhythmic under alternating light-dark conditions, were arrhythmic in complete darkness.

In the larval axolotls, the mediating pathways between the cellular clocks and the indicator process, locomotion, apparently involve hormones. Factors of the pituitary or secretions of one or more of its target organs would seem to be operative as regulators of rhythmic activity. However, the situation in adult axolotls is different. Kalmus detected no activity cycles in them. What changes occur in the mediating pathways or in other parts of the axolotl's timing mechanism as metamorphosis proceeds? Students of developmental biology and endocrinology could help us answer that question.

Whereas adult axolotls and fire salamanders were found to be essentially arrhythmic, adult red backed salamanders, *Plethodon cinereus*, did exhibit cyclic activity under laboratory as well as under field conditions (Ralph, 1957b). Those animals were collected from a cool woodland in southern Michigan early in May, and their motor activity was recorded through the succeeding lunar-month while they were exposed to alternating light (6:00 AM to 6:00 PM) and darkness (6:00 PM to 6:00 AM) or while in constant darkness. Under both sets of conditions, the salamanders were more active during the nighttime than during the daytime (Fig. 42). Maxima were recorded shortly before midnight and minima occurred near midday.

In their natural environments, the red backed salamanders

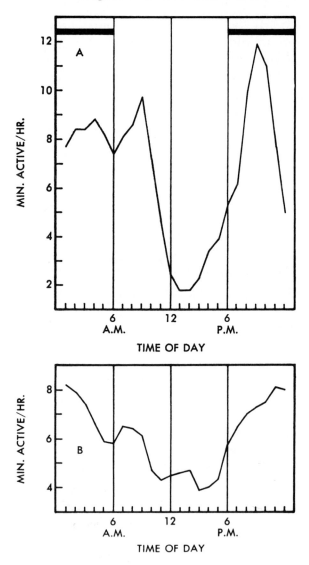

Figure 42. *A.* An average solar-day cycle of activity of red backed sala-manders exposed to alternating light and dark periods. The black bars indicate the hours during which the laboratory was dark. *B.* An average solar-day cycle of activity of red backed salamanders maintained in constant darkness. From: Ralph, C.L.: A diurnal activity rhythm in *Plethodon cinereus* and its modification by an influence having a lunar frequency. *Biol Bull, 113*: 188-197, 1957b.

passed much of the day quietly, sitting under rocks or logs. They became active shortly before 9:00 PM, and crawled about the woodland floor. Hence, the phases of their solar cycles expressed both in the laboratory and in the field were similar, and are highly adaptive to animals which have body coverings through which water moves rapidly. By moving about during the cool dampness of the nighttime, and by hiding under natural cover during the warmer, drier daytime, red backed salamanders are able to keep water losses at minimal levels.

In addition to the circadian rhythm recorded in the laboratory for *Plethodon cinereus*, a lunar one was obvious. The hours of lunar zenith were correlated with depressions in the levels of the salamanders' locomotion. Such depressions were especially obvious within solar-days during which lunar zenith occurred at night. Is it possible that the "light of the moon" causes salamanders to remain in their daytime hiding places? Perhaps, there, they are protected against phase-shifting by light via the depressing effect of lunar zenith itself. Darwin's hypothesis (Chapter 1), suggesting that the sleep position of the leaves of plants inhibits their absorbing moonlight which might shift the phases of their circadian cycles inappropriately, might be extended to these salamanders. It is known that light-dark perturbations effect phase shifts in the daily cycles of activity of some other salamanders, *Plethodon glutinosus*, animals very closely related to the red backed ones (Adler, 1969).

Plethodon glutinosus, commonly called slimy salamanders, were observed in their typical natural habitats in Virginia and North Carolina. There, they, like the red backed salamanders in Michigan, were seen to be nocturnal, leaving their burrows at dusk and returning at dawn. Under laboratory conditions of L/D 9/15 and constant temperature, *Plethodon glutinosus* initiated its activity about the time of the change from light to dark, or about subjective dusk. That phasing was also maintained during two 24-hour periods of darkness which succeeded three days of alternating light and darkness. When the actual times of the changes from light to dark were shifted to later hours of the real solar-day, the commencement of the salamanders' locomotory activity was shifted likewise (Fig. 43). Such new phasing was also retained while the animals lived in constant darkness.

Initiation of activity around the time of previous lights-off, shifting of that phase by light to dark changes and maintenance of the circadian cycles during several days of darkness were recorded for eyeless salamanders as well as for normal ones. In fact, after 20 days of recording—some during L/D regimens and the remainder during darkness—the differences between the performances of eyed and eyeless animals were negligible. Immediately, one questions the manner in which light cues could entrain the cycles of eyeless animals as they did those of normal salamanders.

On the basis of the results of several series of experiments, Adler was able to implicate an extraoptic photoreceptor, probably the forebrain or the pineal body, itself, in the phasing of the activity rhythms of *Plethodon glutinosus*. A similar story is emerging from investigations of the control of circadian cycles in the tiger salamander, *Ambystoma tigrinum* (Taylor, 1970). That eyes are not the only animal receptors which can be stimulated by visible light has been acknowledged for more than 100 years. (See reviews: Adler, 1970 and Steven, 1963).

Among the vertebrates, the forebrain and parts of the pineal system, which develops from the dorsal portion of the midbrain, are known to be extraoptic light receptors. Recently, as for the several amphibians, parts of the pineal system, which Descartes proposed as the seat of the soul, have been proved to be involved in the persistent rhythmicity of mammals (Axelrod, 1970), birds and reptiles (Menaker, 1972). Analogous situations have been reported for a few invertebrates. A circadian rhythm of oviposition can be entrained in normal and in sightless grasshoppers by various light-dark changes (Loher and Chandrashekaran, 1970). In another orthopteran, Dumortier (1972) entrained daily rhythms of stridulation by exposing the animals to L/D regimens. The results were the same whether the insects were normal, had had their compound eyes and ocelli removed, had sustained lesions in the optic lobes of the brain or had had both optic tracts severed.

In addition to finding evidence that an extraoptic photore-

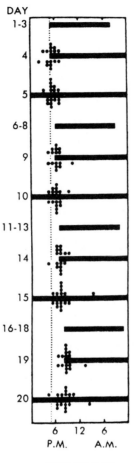

DAY

TIME OF DAY

Figure 43. Activity patterns of normal slimy salamanders following several different L/D regimens. The times of the initiation of the activity of individuals are indicated by the black dots. The black bars show the periods of complete darkness. The vertical dotted line at 5:00 PM is given as a reference point against which to note the variations in the timing of light to dark changes and the corresponding changes in the times of the commencement of the animals' locomotion. From: Adler, K.: Extraoptic phase shifting of circadian locomotor rhythm in salamanders. *Science, 164*: 1290-1292, Fig. 1, 13, June, 1969. Copyright 1969 by the American Association for the Advancement of Science.

ceptor can operate in the phasing of the cycles of the slimy salamanders, Adler (1970) proved the role of a related receptor in the entrainment of activity rhythms of blinded green frogs, *Rana clamitans*. The sightless animals and their normal controls started moving about at the time of lights-on when they were exposed to L/D 14/10. Cycles of circadian frequency continued in both groups of frogs during the several days of constant darkness which followed the period of entrainment.

In this species, the extraoptic light receptor is not the pineal body proper, but is the extracranial pineal end organ or *Stirnorgan,* located in the skin between the frogs' eyes. A Stirnorgan is typical of frogs and toads, but is not found in salamanders. When that organ is removed from blinded green frogs, their activity cycles are no longer entrained by light cues. However, if the intracranial pineal, or the pineal gland proper, is shielded from light by aluminum foil, the rhythms of locomotion of sightless frogs continue to be phased by dark to light changes. As was pointed out earlier, Adler indicated the intracranial pineal body and the forebrain as the most likely receptors of light stimuli which set the time of the beginning of activity in the circadian cycles of blinded slimy salamanders. When these organs were covered by opaque paint, entrainment in response to light-dark changes no longer occurred (Adler, 1970).

Another salamander that can be added to our list of tailed amphibians whose living clocks are indicated by persistent cycles of locomotory activity is the spotted newt, *Notophthalmus (Diemyctylus, Triturus) viridescens.* Adult newts, which are aquatic, were also more active during the night than during the day when kept in the laboratory under constant low illumination (Bennett and Staley, 1960). We collected the newts from ponds in the Blue Ridge region of Virginia where the recordings of their locomotion were made, and we were amazed to see how closely the activity cycles of *Notophthalmus* resembled those of the red backed salamanders, *Plethodon cinereus,* collected in the woodlands of Michigan and studied under constant laboratory conditions in Illinois.

Notophthalmus was maximally active about 1:00 AM (Fig.

44); *Plethodon* was also maximally active about that time (Fig. 42). The newts moved least around midday; the red backed salamanders' minimal point was 3:00 PM, but activity was generally low from 11:00 AM through 4:00 PM. Superimposed upon the 24-hour rhythm of *Notophthalmus* was the same obvious lunar depression of locomotion which Ralph found for *Plethodon*, i.e., hours of lunar zenith correlated with times of lower-than-average levels of activity. These two species of urodeles, so different ecologically, are certainly similar in the phasing of their cycles of motor activity.

Some anurans also have activity rhythms which persist for several days in the laboratory, and whose phasing correlates with the animals' habits in their natural surroundings. The data which Adler has assembled suggest that the green frog, *Rana clamitans*, is day active. An earlier report (Higginbotham, 1939) described the cycles of two species of toads, *Bufo americanus americanus* and *Bufo fowleri*, both of which are obviously night active in their eastern United States environments. Whether the toads lived in constant darkness, under constant light, in a low light intensity

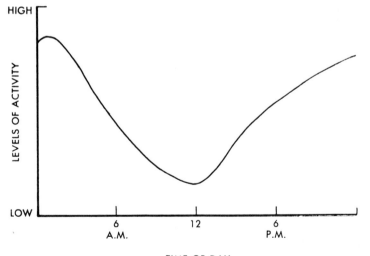

TIME OF DAY

Figure 44. Variations in the levels of activity of spotted newts maintained under constant laboratory conditions.

during the day and in darkness at night or even in darkness by day and in the light by night, they moved about primarily at night. Generally, their peaks of activity occurred near 6:00 PM. The frequency of the toads' cycles was shown to be temperature-compensated, for when the animals' ambient temperatures were doubled or tripled, they continued to vary their levels of locomotion with circadian periodicity.

The frequency of the cycle of activity of another toad, *Bufo regularis,* was neither lengthened nor shortened from 24 hours in constant darkness or in constant light (Cloudsley-Thompson, 1969), and therefore appears to be another exception to Aschoff's Rule that dark active animals have a longer circadian period in the light than in the dark. *Bufo regularis,* nocturnal in its natural habitats of Tanzania, was likewise night active under experimental conditions.

Cycles of oxygen-consumption have also taught us something of the biology of the clocks of Amphibia. Average solar-day and lunar-day rhythms have been described for *Notophthalmus* (Brown, Webb and Bennett, 1958). Cycling metabolic rates were recorded for newts which had been collected from mountain ponds in western Massachusetts and then shipped to Illinois. These cycles shared only a few similarities with the newts' cycles of locomotion. The 24-hour cycle of locomotion, illustrated in Figure 44, is characterized by greater nighttime than daytime activity. The mean daily cycle of oxygen-consumption was of very low amplitude with four peaks of activity distributed fairly evenly in the four quarters of the solar-day (Fig. 45). The peak of the 6:00 AM to noon period was the greatest. The average lunar cycle of metabolism was characterized by minima occurring just before the hours of zenith and nadir of the moon. The same depressing effect of lunar zenith on locomotion has already been emphasized.

The cycles of oxygen-consumption correlated with concurrent changes in the levels of two different geophysical factors, barometric pressure and cosmic radiation. The correlations were virtually identical with those found for fiddler crabs' metabolism and the external factors (Chapter 3) and to those for clams' and oysters' shell movements and the same geophysical changes (Chapter 6). Thus, we have another animal, the spotted newt, in which oxygen-consumption tended to increase as the barometric pressure fell, and tended to fall as the pressure rose.

The results of comparisons of the cycles of the salamanders' metabolism with those of the nucleonic component of radiation illustrated the now familiar partially inverted relationship (Fig. 45). It should be emphasized once more that such correlations, no matter how significant statistically, do not prove that organismic timepieces are paced or regulated directly by changes in the geophysical factors with which the organismic cycles correlate.

However, such relationships do lend credence to the concept that organic and inorganic systems that fluctuate at solar and lunar frequencies may share common regulators or modulators, *e.g.*, geomagnetic or electrostatic fluxes. Before we can dismiss this idea completely, we must prove that our organisms which are functioning with persistent rhythmicity under "constant laboratory conditions" are actually living in a milieu which does

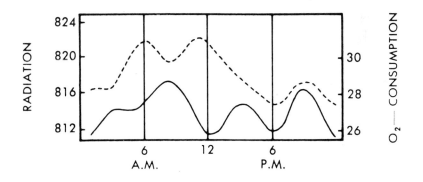

TIME OF DAY

Figure 45.   A 3-hour moving mean of the average daily cycle of the nucleonic component of cosmic radiation for May 12 — June 9, 1954 (broken line) and the 3-hour moving mean of the average daily cycle of oxygen-consumption of spotted newts for the same period (solid line). From: Brown, F.A., Jr., Webb, H.M. and Bennett, M.F.: Comparisons of some fluctuations in cosmic radiation and in organismic activity during 1954, 1955 and 1956. *Am J Physiol, 195*: 237-243, 1958.

not vary, and thereby does not provide cues signaling the time of day, month or year.

In the laboratory, the tropical toad, *Bufo marinus*, consumes oxygen rhythmically when photoperiodic signals are provided (Hutchison and Kohl, 1971). Minimal rates of metabolism were recorded near the time of the onset of light, and maximal rates occurred at the beginning of darkness in L/D regimens of 12/12, 16/8, 6/6 and 23/1. However, when the toads lived in constant light or in constant darkness, no rhythms were apparent. The rhythms that were obvious in L/D neither correlated with parameters of barometric pressure nor with the daily activity cycles of the toads.

But, longer-term variations in the metabolism of one of our best-known laboratory frogs, *Rana pipiens*, do correlate with phases of their annual cycles of activity (Fromm and Johnson, 1955). The seasonal differences reported by these investigators may be the ones that underlie variations in the responses of "summer" and "winter" frogs to identical experimental procedures. All the frogs which Fromm and Johnson studied came into their laboratory in one shipment. The animals were then divided into four groups, *viz.,* 1) those kept under constant illumination at 22° to 28° C and fed at least two weeks before observation; 2) those kept as group one, except without food; 3) those kept as group one, except at 4° C; 4) those kept as group three, but without food.

During a complete year, the oxygen-consumption and the carbon dioxide production of frogs of each group were measured, and their respiratory quotients (R. Q., or the ratio between carbon dioxide produced and oxygen used) were calculated. Similar variations in oxygen-consumption were found for all groups of animals. Maximal rates occurred in the spring, and the minimal ones, which were 40 per cent lower than the maxima, were recorded in the winter. Summer rates were approximately 15 per cent lower than those of the spring, while fall rates were only slightly lower than the summer ones.

The production of carbon dioxide also varied from season to season in like manners in all the groups of frogs. Consequently, differences in respiratory quotients for the four groups paralleled one another. Feeding had no effect on metabolism during any time of the year. The differences in the ambient temperature affected only the levels of oxygen-consumption, and thereby, some of the respiratory quotients.

The results of this particular series of observations supported those of earlier and less precise studies of yearly variations in laboratory frogs, and certainly relate positively to the annual changes in the physiology of *Rana pipiens* in their natural surroundings.

The results of newer investigations of seasonal differences in the same species parallel these physiological variations, too. In spring-summer frogs, urea production was increased after treatment with hyperosmotic sodium chloride solutions for 24 hours; in fall-winter frogs, identical treatment did not evoke such an increase (Jungreis, 1971). In addition, energy reserves change with the season in frogs kept in the laboratory (Mizell, 1965). While blood glucose and fat-body weights were highest in the summer, liver glycogen was highest in the fall.

Diurnal variations in metabolic phenomena of *Rana pipiens* have also been reported, and should be kept in mind by the countless experimentalists who work with this popular frog. K. Upchurch, working in my laboratory, found that the dose of acetylcholine which slowed the heart beat of this species by 12 per cent at midday decreased the cardiac rate by only 6 per cent in the evening. Morgan and Mizell (1971a and 1971b) reported 24-hour rhythms of the synthesis of DNA in two different cell types of *Rana pipiens*. These latter investigators have suggested that lighting regimes regulate the daily cycles of this synthesis. Daily rhythms of heat resistance are known for three other frogs, all hylids, or peepers (Dunlap, 1969; Johnson, 1971 and 1972). In all these species, maximal heat tolerance was measured near midday.

A midday peak in daily cycles of the effectiveness of prolactin, a hormone of the anterior pituitary gland, is typical for a series of vertebrates including *Rana pipiens* (Meier, 1969). Meier and his associates are investigating differences in the responses of various fish, amphibians, reptiles and birds to hormones administered at different times of the solar-day, and have found for all the animals tested that there is a "right time" for treatment. The frogs with which Meier worked lived under natural photoperiods and temperatures. Prolactin was injected on six successive days either at midday or about two hours after sunrise. The animals were sacrificed on the seventh day when their carcasses were assayed for water, lipids and dry nonlipid materials. Midday administration of the hormone induced fattening as indicated by a 200 per cent increase in the levels of lipids in the experimental frogs as compared with those levels in the controls. Treatment with the same amount of hormone in the early morning did not.

More recently, temporal variations in the effects of prolactin on the urodele, *Notophthalmus viridescens,* were reported (Meier, Garcia and Joseph, 1971). In this species, the phasing of the cycle was different from that of the frogs' rhythm. Injections of the hormone late in a 16-hour photoperiod had greater effects on the newts than did administration of prolactin early in the morning or at midday. Other results of Meier's studies (Meier, John and Joseph, 1971) prove that the time of injection of another hormone, corticosterone, sets the phases of the newts' cycles of responsiveness to the prolactin. For example, in *Notophthalmus,* prolactin is most effective eight hours after and is least effective 16 hours after the injection of the adrenal corticoid. In the normal economy of newts, is the response (the water drive, to be discussed later in this chapter) to prolactin timed by the fluctuations in the levels of corticosterone, or by the photoperiod or by a combination of both? Do additional hormones mediate amphibian cycles?

In progress in my laboratory at the present time are investigations of the effects of hormones and physiological stress on the distribution of white blood cells in the peripheral blood of anurans and urodeles. We have sampled blood from *Notophthalmus,* maintained in the dark and at a constant temperature, every three hours through solar periods. The differential counts of the newts' leucocytes were always very much the same. Yet, injections of adrenocorticotrophin (ACTH, another hormone of the anterior pituitary) given at 2:00 AM, 9:00 AM or 2:00 PM caused much greater increases in the numbers of one type of blood cell, the neutrophil, than did administration of the same amount of ACTH at 9:00 PM. What link in the series of reactions that evokes neutrophilia in response to the corticotrophin is so different around 9:00 PM?

Is it adaptive to *Notophthalmus* to react by moving into the water only when the increases in prolactin occur late in the photoperiod? Increases in the titers of that hormone are now known to initiate the water drive in the land phase of the newt (Grant and Grant, 1958). As was pointed out earlier, adult spotted newts are permanently aquatic. The larval stages are also passed in fresh water. However, after metamorphosis and before sexual maturity, a period of three to four years, the animals live on land in the red eft form. As they reach full maturation, their colors change, they acquire smooth, moist skins and they migrate into ponds, streams and lakes.

Reinke and Chadwick (1940) implicated the pituitary gland in the water drive, for implants of that gland from adult newts caused thyroidectomized, gonadectomized and thyroidectomized-gonadectomized efts to go into water in a maximum of eight days. Efts which did not receive implants of pituitaries did not move into the water. The Grants have reviewed the investigations leading to their positive identification of prolactin as the mediator of the water drive, and have emphasized, correctly, how complex a phenomenon the entire water drive syndrome is, hormonally speaking. The reaction is also complicated, temporally speaking.

Are the three to four years between the transformation to the eft stage and full sexual maturity paced off by a biological timing mechanism? We have shown that adult newts, at least, function in time with sun and moon, and perhaps persistent annual cycles result from the combination of the solar and lunar periodicities. Annual variations in the regeneration of forelimbs of *Notophthalmus* are known to persist under controlled conditions (Schauble, 1972). If efts were kept under constant laboratory conditions, would the water drive occur as it does in individuals that are exposed to the many varying factors of our physical world, or, would the young salamanders that lived in the laboratory not secrete increased amounts of prolactin at the usual time?

Another unknown of the water drive of *Notophthalmus* is the manner in which the maturing efts find their ways to water. Even generally, we know very little about the mechanisms of orientation in any of the Amphibia which migrate or home during their life cycles. The great gentleman of salamander biology, the late Victor C. Twitty, proved that the California newt, *Taricha rivularis,* homes to its original stream even when displaced eight miles over mountains (Twitty, 1966). He was not sure of the types of stimuli which provided the necessary cues for the newts. His evidence for olfactory and that for visual guidance were contradictory, and he wondered whether the problem should not have been turned over to the Stanford University chaplain.

Meanwhile, investigators outside the schools of divinity are continuing to try to explain the orientation of amphibians. In field studies of *Taricha granulosa,* the rough skinned newt, no evidence for olfactory guidance was found (Landreth and Ferguson, 1967), but the idea of this newt's relying on a time-compensated sun compass mechanism, similar to that of the honey bee (Chapter 4), was supported by the results of experiments. The salamanders could not orient correctly under cloudy skies, after seven days of complete darkness or after exposures to abnormal L/D regimens. When they were exposed to normal photoperiods and were moving under sunny skies, the animals stayed on course. This type of behavior was true for normal as well as for eyeless newts. What extraoptic photoreceptors can this amphibian use in its orientation? The optic tectum was suggested, for blinded animals from which the optic tecta had been removed, could not orient correctly.

Madison and Shoop (1970) have described the homing behavior of *Plethodon jordani,* a nocturnally active salamander which lives in mixed hardwood-hemlock forests of southeastern United States. The animals were tagged with radioactive wires, and could be traced as they moved about. Observations of the salamanders' climbing on vegetation after displacement and of the positions of their heads suggested olfactory guidance to the investigators. And, since the salamanders homed at night even after having been displaced during the night in opaque containers, and since blinded individuals oriented normally, it was concluded that a celestial compass mechanism was unlikely. Nevertheless, it has not been disproved that *Plethodon jordani* relies upon the position of the moon or formations of stars as points in its compass, and compensates for temporal changes in their positions by using a living clock.

Nocturnal, time-compensated orientation has been claimed for beach fleas (Papi and Pardi, 1963) and for some birds (Sauer and Sauer, 1960). The blinded newts which Madison and Shoop followed may have been able to rely upon extraoptic photoreceptors as they moved through the woods. Extraoptic photoreceptors can function in the orientation of cricket frogs, *Acris gryllus*, (Taylor and Ferguson, 1970) as well as in rough skinned newts, *Taricha granulosa*. However, not only do the cues and receptors of amphibian orientation remain to be worked out, but also its integrating and mediating systems are left to be explained. Certainly, the bits of evidence at hand hint that the living clocks of migrating frogs, toads and salamanders insure that these animals move to the right place at the right time.

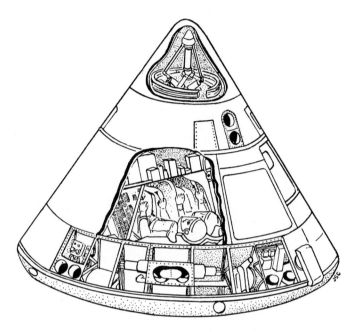

Chapter 8

LIVING CLOCKS—THE FUTURE

"Is it time to wind up The Biological Clock?" That was the
title of a lead article in *Nature New Biology* of May 26, 1971.
The major question posed in that paper was: why is so little
known about the mechanism of the biological clock? The
author(s) suggested that the biological clock remains so elusive
because, ". . . few researchers have been prepared to attempt to
penetrate its secrets." Perhaps a more valid explanation is that not
many investigators have been ready to penetrate *their* secrets.
As emphasized in Chapter 1, timing mechanisms are believed
to be ubiquitous in the living world. Should we not also emphasize that these mechanisms vary to a greater degree than

was anticipated by their earlier investigators or has been acknowledged by present-day researchers?

Organismic timepieces represent a spectrum of phenomena, and a major exercise of the future is the winding up of *these* clocks, the penetration of *their* secrets. A comparative approach, discussed several times in preceding pages, constitutes an especially effective avenue to solutions of these problems. Observations and experiments on the cellular clockworks, the mediating pathways and the hands of the clocks of numerous species—microorganisms, plants and animals—give a researcher the experience necessary to deal realistically with fundamental questions of biochronometry.

The ideas of Pfeffer and Stoppel, now about 100 years old, are still our primary explanations for the workings of living clocks, and most current beliefs are focused upon these concepts. The fully endogenous, self-sustaining, inherited organismic clock, envisioned by Pfeffer, is supported by the majority of present-day students of biological timing. The well analyzed cycles of locomotor activity of cockroaches and the often studied rhythm of hatching of fruit flies are examples of persistent cycles which can be explained in part by Pfeffer's theory.

How are the adherents of this theory attempting to wind up their clock? This clock as a self-sustaining oscillating system was discussed by Bünning in the first edition of his book on biological timing (1958). The mechanism of the hypothetical endogenous chronometer was also likened to an autonomous oscillator or oscillator system by Pittendrigh and Bruce (1957 and 1959). The results of studies by Bünning (1931) on leaf movements in bean plants, by Brown and Webb (1949) on color changes in *Uca*, by Pittendrigh (1954) on eclosion rhythms in *Drosophila* and by Harker (1956) on activity cycles of cockroaches fit neatly into an oscillator model, and were used in the development of that concept. The major characteristics of circadian clocks, free-running periods differing from 24.0 hours, drug- and temperature-compensation and the capacity to be entrained by environmental periodicities, can be accommodated by the oscillator theory.

The 1960 Cold Spring Harbor Symposium on *Biological*

Clocks included at least eight papers addressed to oscillator systems as organismic timing mechanisms. The 1964 Feldafing Summer School on *Circadian Clocks* featured another eight formal discussions on the topic. During the 1969 Friday Harbor Symposium on *Biochronometry,* four papers focused on oscillator systems and clocks were delivered. Two of the latter (Winfree, 1971 and Pavlidis, 1971) described oscillator models—mathematical, physical and biochemical—which could function heuristically in penetrating the secrets of biological clocks. The development and testing of such models are in progress, and represent a major attack on the solution of the mechanism of living clocks for the future. Toward accomplishment of that end, investigations of the reality of the chronon and studies of changes in enzyme systems are important pathways.

Discussions during the 1971 Tihany Symposium on *Invertebrate Neurobiology and the Mechanisms of Rhythm Regulation* also pointed up systems and methods to be exploited in our future investigations of biochronometry. Pacemaker activities of molluscan neurons and of crustacean ganglia were described. Cyclic actions of invertebrate hearts were analyzed. Neural networks whose rhythmic driving actions account for behavioral acts of molluscs, annelids and arthropods were discussed. Finally, regulation of longer-term cycles—circadian and tidal—were considered.

Throughout that range of discussions, several questions stood out: what similarities do these diverse, yet related, phenomena share? What are their major differences? One could not escape realizing once again that the mechanisms of timing vary even in the animal kingdom, but that functions common to all cells, membrane permeability, ion distribution, production of electric potentials and currents, bioenergetics, syntheses and secretion, must be understood in detail before we will be able to explain completely the rhythmic activities of living systems.

Future studies of the many clocks of organisms can be expected to elucidate further basic cellular phenomena, and to relate them to biological timekeeping. Incidental to defining these relationships, we may help bridge the gap which is said to exist between molecular and organismic biology.

Investigations at the cellular and organismic levels and the development and testing of models will also be involved in penetrating the secrets of exogenously modulated biological clocks, those of the type originally postulated by Stoppel. I have emphasized that the theory that living chronometers are paced or kept in time with their physical surroundings by cyclic variations in geophysical factors, and are dependent upon such changes for frequency maintenance, has been developed and defended most successfully by Brown, one of the very few supporters of Stoppel's ideas.

Brown's papers of the Cold Spring Harbor Symposium of 1960, the Feldafing Summer School of 1965 and in *The Biological Clock: Two Views* review his evidence for the existence of environmentally dependent organismic timing systems. The average cycles of metabolism of potato tubers (Brown, 1958 and 1960), of oxygen-consumption in fiddler crabs, of shell movements of molluscs and of the orientation of mud snails are some of the rhythms which best illustrate the manifestations of living clocks held to depend upon subtle geophysical factors in their environments.

The correlations between cycles of animal metabolism and cycles of barometric pressure and cosmic radiation, cited in this book, have been discussed at length by Brown in support of his explanation of biological clocks. Also noted in my discussions were the investigations of the effects of variations in magnetic and electrostatic forces on rhythmically functioning organisms— honey bees, earthworms and snails. In addition, studies of geomagnetism and navigation of birds were cited. Some of the results of the investigation of this ornithological problem have strengthened Brown's claims and his concept of the Space-Time Continuum as a matrix which gives organisms information about their temporal and spatial situations.

A very real hurdle to the acceptance of the reality of exogenously regulated organismic clocks is the observation that under laboratory conditions of constant light and temperature, many animal and plant rhythms free-run, *i.e.*, as explained in Chapter 1, their frequencies are slightly different from those which obtain when the organisms live under natural conditions or under entraining regimens. Free-running circadian cycles are not exactly 24.0 hours; free-running tidal cycles are not exactly 12.4 hours. If organismic rhythms do depend upon external cues that come at precise geophysical frequencies, how can free-running be resolved?

Brown's theory of autophasing (1959 and 1972) attempts to explain free-running in terms of phase-shifting, a phenomenon well-known and one amenable to further observation and experimentation. In Chapter 2, fiddler crabs' cycles of sensitivity to light and temperature were described. During a 24-hour period, the crabs' responses to perturbations of the same intensity vary, and do so because of the animals' own rhythms of sensitivity. These rhythms are the ones generated, according to Brown, by the interaction of the living organisms and unknown, subtle geophysical factors.

Elegant analyses by Pittendrigh and his associates (*e.g.*, Pittendrigh and Minis, 1971) of phase-response relationships reinforce the idea of temporally varying reactions to identical stimuli. This flexibility is thought to be highly beneficial to rhythmic forms in their natural cyclically changing environments, for by virtue of their differing levels of sensitivity, the organisms can set the phases of their own rhythms with specific events—sunrise, sunset, low tide—in their surroundings in adaptive manners, and in so doing can also function at precise solar and lunar frequencies.

However, again according to Brown's theory, when the living systems are placed under constant conditions, they use their own rhythmic fluctuations to shift phases of their cycles, as follows: although light and temperature are kept physically constant, the influences of these factors on the organisms are greater or lesser depending upon the organisms' levels of sensitivity. When the levels are high, the constant light or temperature is a perturbation, and serves to shift the organismic phases and thereby, changes the periods of the cycles to more or less than 24.0, 12.4 or 24.8 hours. When the organisms are in insensitive phases of their cycles, the constant light or temperature does not cause a shift in phasing and during these times, probably has no effect on period length.

Obvious questions to be answered by future investigators of environmentally dependent living clocks are these: which geophysical factors pace the clocks? What are the organisms' receptors of the subtle geophysical changes? How do the physical factors and cells interact in their generation of the organismic cycles of sensitivity? How do changes in light intensity or temperature cause phase-shifting and consequently, free-running?

And, how can we prove whether Pfeffer or Stoppel was correct? How can we prove whether cells are constantly dependent upon their external environments for their timekeeping? These questions have been asked often in the past, are being asked now and will be asked in the future. Recording of processes, which are rhythmic on earth, *e.g.*, hatching of fruit flies, oxygen-consumption of potatoes, locomotion of cockroaches, from systems living in space capsules or in space laboratories, has been considered seriously, and some of its instruments have even been designed (Lear, 1965).

Our reasoning holds that the organisms in space would be away from the influences of geophysical cycles typical of our earth; the plants or animals would be deprived of information signaling time on earth. If under those conditions, the organisms were to continue to function rhythmically as they do on earth, one might conclude that their clocks are indeed endogenous.

If, on the other hand, the organisms no longer behaved rhythmically or if their cycles were aberrant, investigators might agree that biological clocks are dependent upon variations in geophysical factors as such changes occur on and near earth. Therefore, an obvious type of experiment for the future has been planned. Its execution depends upon the technological, the economic and probably also the political atmosphere of the future.

If that experiment is set up and if its results are transmitted successfully to students of biochronometry waiting on earth or standing by in space laboratories, what will have been accomplished? Will we really have any more reason than we have now to postulate that living clocks are all the same, especially if we test only one or two specific types in space? A spectrum of biochronometers would continue to be plausible, and the secrets of all conceivable kinds of living clocks would still beg clarification. With the evidence at hand at present, it is impossible to defend the view that all biological clocks are *either* endogenous *or* environmentally dependent ones. Some cells appear to beat out time completely independent of their environments; some others appear to demand constant information from their surroundings to pace off their frequencies; still others probably lie between the two extremes. Investigations on earth can help penetrate some of the secrets of all types.

Among the secrets to be investigated in the future are two which appear now to be characteristics of most of the biological clocks we know. These are the clocks' relative insensitivity to drugs and their temperature-independence or temperature-compensation. Cited in previous chapters were the results of recent investigations which indicate that some unidentified parts of organismic clockworks are affected by heavy water, ethyl alcohol or antibiotics.

These findings focus attention once more on relationships between activities common and necessary to all cells, *e.g.*, maintenance of normal membrane permeability, distribution of ions, and syntheses of specific proteins as directed by DNA and RNA, and timing mechanisms. These findings also cause us to reconsider the conclusions of older studies about drugs and biological clocks, mentioned earlier. Did the thyroid extract or the quinine or the cyanide tested get to the sensitive gears of the clocks? Are reactions known to be sensitive to these drugs actually working parts of biological clocks? Whatever the final decisions are, all these results, old and new, impress us by underlining the superb homeostatic properties of living clocks, properties which will be explained fully by future investigations.

Temperature-independence or temperature-compensation of period length is no longer so mysterious an attribute of organismic clocks as it seemed to be even 15 years ago, and biochemical studies of the present and near future should allow us to explain this highly adaptive characteristic, too. The Q_{10}'s of most rhythms are very close to 1.0 (Sweeney and Hastings, 1960). That point has been appreciated for almost 100 years, and has been one of the facts which adherents of the theory of the environmentally dependent clock found in support of their ideas. They asked how a completely endogenous metabolic set of clockworks could maintain its usual frequency at temperatures so different as 5° and 25° C. They reasoned that clocks dependent upon geophysical factors could function that way, because they are paced by changes in the environment which occur no matter what its temperature may be. But, metabolically dependent or completely endogenous timepieces could not maintain constant frequencies in the face of temperature change, for most biochemical reactions are increased two- to four-fold for a 10° C increase in temperature.

However, during the last 15 years, it has become more and more apparent that not all metabolic reactions have Q_{10}'s lying in the classically stated range of 2 to 4. Many are temperature-compensated in their so-called physiological ranges. Consequently, we realized that temperature-compensation is not unique to biological clocks, and need not be a barrier to the penetration of the metabolic secrets of these clocks.

In working out temperature-compensation of living chronometers, we will go to various structures and many different reactions at the cellular level. Burkhardt (1959) discovered temperature-compensation, and therefore Q_{10}'s close to 1.0, for stretch receptors of crayfish. The frequency of impulses from those cells was much the same through a range of temperatures. Burkhardt believes that the nerve cells, themselves, compensate by varying their generator potentials and thresholds as the temperature changes. Membranes and bioelectrical activities are implicated in the compensatory reactions.

Newell (1969) found temperature-independent metabolism in several intertidal invertebrates, amphipods, limpets and cockles, and Percy and Aldrich (1971) reported that the metabolic rates of tissues of oysters are temperature-independent in the range, 12° to 40° C. Through the temperature range typical of their native habitats, broad-leafed cattails show thermal insensitivity (McNaughton, 1972). Changes in enzymes, enzyme systems, respiratory reactions and adjustments of mitochondrial activities have all been brought into the explanation of temperature-compensation (Brinkmann, 1971), and are reactions which invite intensive analyses in our future work on biological clocks.

Other analyses of organismic timing will focus upon phase-shifting and its agents. Descriptions and the results of many such studies accomplished to date have been included in preceding chapters. As was suggested in this chapter, the actual mechanism of shifting has not as yet been determined. Brown and Webb (1949) and Pittendrigh (1960) formulated hypothetical, partial explanations, both of which are based on two-component systems. A basic center or oscillator (the clockworks?) is said to maintain its frequency constantly, but can be set in various temporal relationships with a second component which determines the timing of specific phases of organisms' cycles, and can be manipulated by Zeitgebers or phase-setting factors.

This two-part model fits many of our facts, but is not complete enough to explain all the actions and reactions which transpire during phase-setting and phase-shifting. The receptors and the afferent pathways operating remain to be cleared up. The manners in which perturbations are integrated and transmitted to setting and shifting actions, themselves unknown, have not been worked out. What types of transduction occur? Then, how and where is the resulting information fed into mediating pathways and the hands of the clocks? Future studies of well-known Zeitgebers such as changes in light intensity, changes in temperature, and of those newly discovered such as sound and social interactions can be expected to answer these questions and to contribute to the winding up of biological clocks.

More detailed analyses of mediating pathways of animals' rhythms are sure to become available. As general endocrinologists and neurobiologists continue to describe the finer points of the anatomy and physiology of glands and neural structures, students of biochronometry should have more of the information needed to clarify the tie-ups between animals' clockworks and the clocks' hands. Precise investigations of nervous pacemakers and neuronal networks and their regulation of particular behavioral acts are giving us working models of possible regulatory pathways. Electron microscopic studies of neurosecretory cells and neurosecretory systems can lead to greater understanding of the chemical mediation of rhythms in animals.

Undoubtedly, the number of specific examples of living clocks known now will be increased many times as studies in biochronometry extend into the future. When more are added to our already lengthy list which spans the living world from microorganisms through the angiosperms in one direction and through the mammals and the arthropods in others, new types of clocks may be recognized. Knowledge of such possible new types might help penetrate the secrets of clocks which have been studied in detail, *e.g.,* those of fiddler crabs, honey bees, fruit flies and cockroaches, but about which much remains unknown.

Greater understanding of all animals' clocks would have potentially great practical applications for man. He may be able to control animal pests and weeds by manipulating their clocks and setting their hands in relationships with their environments which are lethal. He may be able to take advantage of his own most efficient phases and increase industrial productivity. By knowing the temporal as well as the spatial environment of the communities he studies, an ecologist may be better prepared to understand and preserve natural groupings and their surroundings. By analyzing his own temporal functioning and its requirements, man in space may be safer than he has been in the past.

Finally, students of adaptations will continue to look at biological clocks in terms of their meaning in the everyday lives and in the evolution of organisms. In this book, I have cited some obvious examples of advantages of the possession of clocks to their possessors. Precise timing of activities aids in the orientation and communication of bees, in the water regulation of earthworms and in the feeding of intertidal animals. Questionable cases, *e.g.,* color change rhythms in fiddler crabs, have also been pointed out. In addition, I suggested that lunar cycles of fresh water and terrestrial organisms are examples of physiological relics which are of no adaptive value to their present-day practitioners. Enright (1970) questions the whole concept of animal clocks' being adaptive and of selective value. An authority on invertebrate physiology has stated bluntly that rhythms really do not make any difference to animals' behaviors. So, even ideas about the value of living timepieces represent a wide spectrum, as do the many biological clocks, themselves.

To wind up these clocks, to penetrate their secrets are problems for the future. Ours may be the most time-conscious generation that has ever lived, but one may very well expect that those that follow will be equally or even more so, and I hope that investigations of biological clocks will continue to be manifestations of that consciousness.

BIBLIOGRAPHY

Abramowitz, A. A.: The chromatophorotropic hormone of the Crustacea; standardization, properties and physiology of the eye-stalk glands. *Biol Bull, 72*: 344-365, 1937.

Adler, K.: Extraoptic phase shifting of circadian locomotor rhythm in salamanders. *Science, 164*: 1290-1292, 1969.

Adler, K.: The role of extraoptic photoreceptors in amphibian rhythms and orientation: A review. *J Herpetol, 4*: 99-122, 1970.

Ajrapetyan, S. N.: On the regulation of rhythmic activity of *Helix* neurons. In Salánki, J. (Ed.): *Invertebrate Neurobiology: Mechanisms of Rhythm Regulation.* Hungarian Academy of Sciences, Budapest, 1973, in press.

Arbit, J.: Diurnal cycles and learning in earthworms. *Science, 126*: 654-655, 1957.

Aréchiga, H., Fuentes, B. and Barrera, B.: On the dynamics of circadian rhythm of activity in the crayfish nervous system. In Salánki, J. (Ed.): *Invertebrate Neurobiology: Mechanisms of Rhythm Regulation.* Hungarian Academy of Sciences, Budapest, 1973, in press.

Arudpragasam, K. D. and Naylor, E.: Gill ventilation volumes, oxygen consumption and respiratory rhythms in *Carcinus maenas* (L.). *J Exp Biol, 41*: 309-321, 1964.

Aschoff, J. (Ed.): *Circadian Clocks,* Proceedings of the Feldafing Summer School, 1964. Amsterdam, North-Holland Publ. Co., 1965, 479 pp.

Aschoff, J.: Exogenous and endogenous components in circadian rhythms. *Cold Spring Harbor Symp Quant Biol, 25*: 11-28, 1960.

Aschoff, J.: The phase-angle difference in circadian periodicity. In Aschoff, J. (Ed.): *Circadian Clocks.* Amsterdam, North-Holland Publ. Co., 1965, pp. 262-276.

Aschoff, J.: Circadian rhythms: Vertebrates and man. *Proc Internat Union Physiol Sci, 8*: 15-16, 1971.

Axelrod, J.: The pineal gland. *Endeavour, 29*: 144-148, 1970.

Baldwin, F. M.: Diurnal activity of the earthworm. *J Anim Behav, 7*: 187-190, 1917.

Barnett, A., Ehret, C. F. and Wille, J. J.: Testing the chronon theory of circadian timekeeping. In Menaker, M. (Ed.): *Biochronometry.* Washington, D. C., National Academy of Sciences, 1971, pp. 637-651.

Barnothy, M. F. (Ed.): *Biological Effects of Magnetic Fields.* New York, Plenum Press, 1964, vol. I, 324 pp.

Barnothy, M. F. (Ed.): *Biological Effects of Magnetic Fields.* New York and London, Plenum Press, 1969, vol. II, 314 pp.

Barnwell, F. H.: Observations on daily and tidal rhythms in some fiddler crabs from equatorial Brazil. *Biol Bull, 125*: 399-415, 1963.

Barnwell, F. H.: Daily and tidal patterns of activity in individual fiddler crabs (genus *Uca*) from the Woods Hole region. *Biol Bull, 130*: 1-17, 1966.

Barnwell, F. H.: Comparative aspects of the chromatophoric responses to light and temperature in fiddler crabs of the genus *Uca*. *Biol Bull, 134*: 221-234, 1968.

Barnwell, F. H.: Overt rhythms in the shell movements of the American oyster. In Scheving, L. E., Halberg, F. and Pauly, J. E. (Eds.): *Chronobiology*. Proceedings of the Society for the Study of Biological Rhythms, Little Rock, Ark., Tokyo, Igaku Shoin, Ltd., in press, 1973.

Barnwell, F. H. and Brown, F. A., Jr.: Differences in the persistent metabolic rhythms of fiddler crabs from two levels of the same beach. *Biol Bull, 125*: 371-372, 1963.

Beck, S. D.: *Insect Photoperiodism*. New York, Academic Press, 1968, 288 pp.

Beier, W.: Beeinflussung der Inneren Uhr der Bienen durch Phasenverschiebung des Licht-Dunkel-Zeitgebers. *Z Bienenforsch, 9*: 356-378, 1968.

Beling, I.: Über das Zeitgedächtnis der Bienen. *Z Vergl Physiol, 9*: 259-338, 1929.

Bennett, M. F.: The rhythmic activity of the quahog, *Venus mercenaria*, and its modification by light. *Biol Bull, 107*: 174-191, 1954.

Bennett, M. F.: The phasing of the cycle of motor activity in the fiddler crab, *Uca pugnax*. *Z Vergl Physiol, 47*: 431-437, 1963.

Bennett, M. F.: Isolation of the brain and rhythmicity in earthworms. *Z Vergl Physiol, 56*: 376-379, 1967.

Bennett, M. F.: Persistent seasonal variations in the diurnal cycle of earthworms. *Z Vergl Physiol, 60*: 34-40, 1968.

Bennett, M. F.: The clock and the calendar of the earthworm. *Med Opinion and Review, 5*: 44-56, 1969.

Bennett, M. F.: Bilateral symmetry in the mediation of circadian differences in earthworms. *Z Vergl Physiol, 69*: 1-5, 1970.

Bennett, M. F.: The central nervous system and circadian differences in the earthworm. In Salánki, J. (Ed.): *Invertebrate Neurobiology: Mechanisms of Rhythm Regulation*. Hungarian Academy of Sciences, Budapest, 1973, in press.

Bennett, M. F. and Brown, F. A., Jr.: Experimental modification of the lunar rhythm of running activity of the fiddler crab, *Uca pugnax*. *Biol Bull, 117*: 404, 1959.

Bennett, M. F. and Guilford, C. B.: Circadian variations in the oxygen-consumption of anterior segments of earthworms. *Z Vergl Physiol, 74*: 32-38, 1971.

Bennett, M. F. and Huguenin, J.: Geomagnetic effects on a circadian difference in reaction times in earthworms. *Z Vergl Physiol, 63:* 440-445, 1969.

Bennett, M. F. and Huguenin, J.: Geomagnetism and circadian organization in earthworms. *Proc Internat Union Physiol Sci, 9:* 53, 1971.

Bennett, M. F. and Reinschmidt, D. C.: The diurnal cycle and a difference in reaction times in earthworms. *Z Vergl Physiol, 49:* 407-411, 1965a.

Bennett, M. F. and Reinschmidt, D. C.: The diurnal cycle and locomotion in earthworms. *Z Vergl Physiol, 51:* 224-226, 1965b.

Bennett, M. F. and Renner, M.: The collecting performance of honey bees under laboratory conditions. *Biol Bull, 125:* 416-430, 1963.

Bennett, M. F., Shriner, J. and Brown, R. A.: Persistent tidal cycles of spontaneous motor activity in the fiddler crab, *Uca pugnax. Biol Bull, 112:* 267-275, 1957.

Bennett, M. F. and Staley, J.: Cycles of motor activity in the newt, *Triturus viridescens. Anat Rec, 137:* 339, 1960.

Bennett, M. F. and Willis, M. H.: The brain and rhythmicity in earthworms. *Z Vergl Physiol, 53:* 95-98, 1966.

Bentley, P. J.: Adaptations of Amphibia to arid environments, *Science, 152:* 619-623, 1966.

Biological Clocks. Cold Spring Harbor Symp Quant Biol, 25: 1960, 524 pp.

Bliss, D.: Neuroendocrine control of locomotor activity in the land crab, *Gecarcinus lateralis. Memoirs of the Society for Endocrinology, 12:* 391-410, Heller, H. and Clark, R. B. (Eds.): London and New York, Academic Press, 1962.

Bohn, G.: Periodicitè vitale des animaux soumis aux oscillations du niveau des hautes mers. *C R Acad Sci* (Paris), *139:* 610-611, 1904.

Brady, J.: How are insect circadian rhythms controlled? *Nature, 223:* 781-784, 1969.

Bräuninger, H. D.: Über den Einfluss meteorologischer Faktoren auf die Entfernungsweisung im Tanz der Bienen. *Z Vergl Physiol, 48:* 1-130, 1964.

Brinkmann, K.: Metabolic control of temperature compensation in the circadian rhythm of *Euglena gracilis.* In Menaker, M. (Ed.): *Biochronometry.* Washington, D. C., National Academy of Sciences, 1971, pp. 567-593.

Brown, F. A., Jr.: Studies on the physiology of *Uca* red chromatophores. *Biol Bull, 98:* 218-226, 1950.

Brown, F. A., Jr.: Biological clocks and the fiddler crab. *Sci Am, 190(4):* 34-37, 1954a.

Brown, F. A., Jr.: Persistent activity rhythms in the oyster. *Am J Physiol, 178:* 510-514, 1954b.

Brown, F. A., Jr.: An exogenous reference-clock for persistent temperature-independent, labile, biological rhythms. *Biol Bull, 115:* 81-100, 1958.

Brown, F. A., Jr.: Living clocks. *Science, 130*: 1534-1544, 1959.

Brown, F. A., Jr.: Response to pervasive geophysical factors and the biological clock problem. *Cold Spring Harbor Symp Quant Biol, 25*: 57-71, 1960.

Brown, F. A., Jr.: Diurnal rhythm in cave crayfish. *Nature, 191*: 929-930, 1961.

Brown, F. A., Jr.: A unified theory for biological rhythms: rhythmic duplicity and the genesis of "circa" periodisms. In Aschoff, J. (Ed.): *Circadian Clocks*. Amsterdam, North-Holland Publ. Co., 1965, pp. 231-261.

Brown, F. A., Jr.: A hypothesis for extrinsic timing of circadian rhythms. *Can J Bot, 47*: 287-298, 1969.

Brown, F. A., Jr.: Hypothesis of environmental timing of the clock. In Palmer, J. D. (Ed.): *The Biological Clock. Two Views*. New York and London, Academic Press, 1970, pp. 13-59.

Brown, F. A., Jr.: The "clocks" timing biological rhythms. *Am Sci, 60*: 756-766, 1972.

Brown, F. A., Jr., Bennett, M. F. and Ralph, C. L.: Apparent reversible influence of cosmic-ray-induced showers upon a biological system. *Proc Soc Exp Biol Med, 89*: 332-337, 1955.

Brown, F. A., Jr., Bennett, M. F. and Webb, H. M.: Persistent daily and tidal rhythms of O_2-consumption in fiddler crabs. *J Cell Comp Physiol, 44*: 477-505, 1954.

Brown, F. A., Jr., Bennett, M. F. and Webb, H. M.: A magnetic compass response of an organism. *Biol Bull, 119*: 65-74, 1960.

Brown, F. A., Jr., Bennett, M. F., Webb, H. M. and Ralph, C. L.: Persistent daily, monthly and 27-day cycles of activity in the oyster and quahog. *J Exp Zool, 131*: 235-262, 1956.

Brown, F. A., Jr., Fingerman, M., Sandeen, M. I. and Webb, H. M.: Persistent diurnal and tidal rhythms of color change in the fiddler crab, *Uca pugnax. J Exp Zool, 123*: 29-60, 1953.

Brown, F. A., Jr. and Park, Y. H.: Duration of an after-effect in planarians following a reversed horizontal magnetic vector. *Biol Bull, 128*: 347-355, 1965a.

Brown, F. A., Jr. and Park, Y. H.: Phase-shifting a lunar rhythm in planarians by altering the horizontal magnetic vector. *Biol Bull, 129*: 79-86, 1965b.

Brown, F. A., Jr. and Webb, H. M.: Temperature relations of an endogenous daily rhythmicity in the fiddler crab, *Uca. Physiol Zool, 21*: 371-381, 1948.

Brown, F. A., Jr. and Webb, H. M.: Studies of the daily rhythmicity of the fiddler crab, *Uca*. Modifications by light. *Physiol Zool, 22*: 136-148, 1949.

Brown, F. A., Jr., Webb, H. M. and Bennett, M. F.: Proof for an endogenous component in persistent solar and lunar rhythmicity in organisms. *Proc Nat Acad Sci, 41*: 93-100, 1955.

Brown, F. A., Jr., Webb, H. M. and Bennett, M. F.: Comparisons of some fluctuations in cosmic radiation and in organismic activity during 1954, 1955 and 1956. *Am J Physiol, 195*: 237-243, 1958.

Brown, F. A., Jr., Webb, H. M., Bennett, M. F. and Sandeen, M. I.: Temperature-independence of the frequency of the endogenous tidal rhythm of *Uca. Physiol Zool, 27*: 345-349, 1954.

Brown, F. A., Jr., Webb, H. M. and Brett, W. J.: Exogenous timing of solar and lunar periodisms in metabolism of the mud snail, *Ilyanassa* (= *Nassarius) obsoleta*, in laboratory constant conditions. *Gunma J Med Sci, 8*: 233-242, 1959.

Bünning, E.: Untersuchungen über die autonomen tagesperiodischen Bewegungen der Primärblätter von *Phaseolus multiflorus. Jb Wiss Bot, 75*: 439-480, 1931.

Bünning, E.: *Die Physiologische Uhr.* Berlin-Göttingen-Heidelberg, Springer-Verlag, 1958, 105 pp.

Bünning, E.: Opening address: Biological clocks. *Cold Spring Harbor Symp Quant Biol, 25*: 1-9, 1960.

Bünning, E.: *The Physiological Clock.* New York, Springer-Verlag, 1967, 167 pp.

Bünning, E.: The adaptive value of circadian leaf movements. In Menaker, M. (Ed.):*Biochronometry.* Washington, D. C., National Academy of Sciences, 1971, pp. 203-211.

Burkhardt, D.: Die Erregungsvorgänge sensibler Ganglienzellen in Abhängigkeit von der Temperatur. *Biol Zentralbl, 78*: 22-62, 1959.

Cloudsley-Thompson, J. L.: *Rhythmic Activity in Animal Physiology and Behaviour.* New York and London, Academic Press, 1961, 236 pp.

Cloudsley-Thompson, J. L.: *The Zoology of Tropical Africa.* New York, W. W. Norton and Co., Inc., 1969, 355 pp.

Cole, L. C.: Biological clock in the unicorn. *Science, 125*: 874-876, 1957.

Conroy, R. T. W. L. and Mills, J. N.: *Human Circadian Rhythms.* London, J. & A. Churchill, 1970, 236 pp.

Crane, J.: On the color changes of fiddler crabs (Genus *Uca*) in the field. *Zoologica, 29*: 161-168, 1944.

Crane, J.: Aspects of social behavior in fiddler crabs, with special reference to *Uca maracoani (Latreille). Zoologica, 43*: 113-130, 1958.

Darwin, C.: *The Formation of Vegetable Mould through the Action of Worms.* London, John Murray, 1881, 326 pp.

DeCoursey, P. J.: Phase control of activity in a rodent. *Cold Spring Harbor Symp Quant Biol, 25*: 49-55, 1960.

Dodd, J. R.: Effect of light on rate of growth of bivalves. *Nature, 224*: 617-618, 1969.

Dumortier, B.: Photoreception in the circadian rhythm of stridulatory activity in *Ephippiger* (Ins., Orthoptera). *J Comp Physiol, 77*: 80-112, 1972.

Dunlap, D. G.: Influence of temperature and duration of acclimation, time of day, sex and body weight on metabolic rates in the hylid frog, *Acris crepitans. Comp Biochem Physiol, 31*: 555-570, 1969.

Ehret, C. F. and Trucco, E.: Molecular models for the circadian clock. I. The chronon concept. *J Theor Biol, 15*: 240-262, 1967.

Eiseley, L.: *The Unexpected Universe.* New York, Harcourt, Brace and World, Inc., 1969, 239 pp.

Emlen, S. T.: The influence of magnetic information on the orientation of the indigo bunting, *Passerina cyanea. Anim Behav, 18*: 215-224, 1970.

Enright, J. T.: The search for rhythmicity in biological time-series. *J Theor Biol, 8*: 426-468, 1965.

Enright, J. T.: Ecological aspects of endogenous rhythmicity. *Ann Rev Ecol System, 1*: 221-238, 1970.

Enright, J. T.: Heavy water slows biological timing processes. *Z Vergl Physiol, 72*: 1-16, 1971a.

Enright, J. T.: The internal clock of drunken isopods. *Z Vergl Physiol, 75*: 332-346, 1971b.

Eskin, A.: Properties of the *Aplysia* visual system: *in vitro* entrainment of the circadian rhythm and centrifugal regulation of the eye. *Z Vergl Physiol, 74*: 353-371, 1971.

Evans, J. W.: Tidal growth increments in the cockle *Clinocardium nuttalli. Science, 176*: 416-417, 1972.

Fingerman, M.: Tidal rhythmicity in marine organisms. *Cold Spring Harbor Symp Quant Biol, 25*: 481-489, 1960.

Fingerman, M.: Crustacean color change with emphasis on the fiddler crab. *Scientia, 103*: 1-16, 1968.

Fingerman, M.: Perspectives in crustacean endocrinology. *Scientia, 105*: 1-23, 1970a.

Fingerman, M.: Comparative physiology: Chromatophores. *Ann Rev Physiol, 32*: 345-372, 1970b.

Fingerman, M. and Lago, A. D.: Endogenous twenty-four hour rhythms of locomotor activity and oxygen consumption in the crawfish *Orconectes clypeatus. Am Midl Nat, 58*: 383-393, 1957.

Forel, A. H.: *Das Sinnesleben der Insekten,* (M. Semon, trans.). München, Ernst Reinhardt Verlag, 1910, 393 pp.

Friedman, H., Becker, R. O. and Bachman, C. H.: Effect of magnetic fields on (human) reaction time performance. *Nature, 213*: 949-950, 1967.

Fromm, P. O. and Johnson, R. E.: The respiratory metabolism of frogs as related to season. *J Cell Comp Physiol, 45*: 343-359, 1955.

Gompel, M.: Recherches sur la consommation d'oxygène de quelques animaux aquatiques littoraux. *C R Acad Sci (Paris), 205*: 816-818, 1937.

Gould, J. L., Henery, M. and MacLeod, M. C.: Communication of direction by the honey bee. *Science, 169*: 544-554, 1970.

Grant, W. C., Jr. and Grant, J. A.: Water drive studies on hypophysectomized efts of *Diemyctylus viridescens. Biol Bull, 114*: 1-9, 1958.

Guyselman, J. B.: Solar and lunar rhythms of locomotor activity in the crayfish *Cambarus virilis. Physiol Zool, 30*: 70-87, 1957.

Gwinner, E.: Periodicity of a circadian rhythm in birds by species-specific song cycles. *Experientia, 22*: 765-766, 1966.

Harker, J. E.: Factors controlling the diurnal rhythm of activity of *Periplaneta americana L. J Exp Biol, 33*: 224-234, 1956.

Harker, J. E.: *The Physiology of Diurnal Rhythms.* Cambridge, Cambridge University Press, 1964, 114 pp.

Hastings, J. W.: Cellular-biochemical clock hypothesis. In Palmer, J. D. (Ed.): *The Biological Clock. Two Views.* New York and London, Academic Press, 1970, pp. 61-91.

Hauenschild, C.: Die Schwärmperiodizität von *Platynereis dumerilii* im DD-LD-Belichtungszyklus und nach Augenausschaltung. *Z Naturforsch, 16b*: 753-756, 1961.

Heusner, A.: Sources of error in the study of diurnal rhythm in energy metabolism. In Aschoff, J. (Ed.): *Circadian Clocks.* Amsterdam, North-Holland Publ. Co., 1965, pp. 3-12.

Higginbotham, A. C.: Studies on amphibian activity. I. Preliminary report on rhythmic activity of *Bufo americanus americanus holbrook* and *Bufo fowleri hinckley. Ecology, 20*: 58-70, 1939.

Hines, M. N.: A tidal rhythm in behavior of melanophores in autotomized legs of *Uca pugnax. Biol Bull, 107*: 386-396, 1954.

Hogben, L. and Slome, D.: The pigmentary effector system. VI. The dual character of endocrine coordination in amphibian colour change. *Proc R Soc Lond, 108B*: 10-53, 1931.

Hutchison, V. H. and Kohl, M. A.: The effect of photoperiod on daily rhythms of oxygen consumption in the tropical toad *Bufo marinus. Z Vergl Physiol, 75*: 367-382, 1971.

Is it time to wind up the biological clock? *Nature New Biology, 231*: 97-98, 1971.

Jacklet, J. W.: Circadian rhythm of optic nerve impulses recorded in darkness from isolated eye of *Aplysia. Science, 164*: 562-563, 1969.

Jacklet, J. W. and Geronimo, J.: Circadian rhythm: Population of interacting neurons. *Science, 174*: 299-302, 1971.

Jegla, T. C. and Poulson, T. L.: Evidence of circadian rhythms in a cave crayfish. *J Exp Zool, 168:* 273-282, 1968.

Jegla, T. C. and Poulson, T. L.: Circannian rhythms-I. Reproduction in the cave crayfish, *Orconectes pellucidus inermis. Comp Biochem Physiol, 33:* 347-355, 1970.

Johnson, C. R.: Daily variation in the thermal tolerance of *Litoria caerulea* (*Anura*: *Hylidae*). *Comp Biochem Physiol, 40A:* 1109-1111, 1971.

Johnson, C. R.: Diel variation in the thermal tolerance of *Litoria gracilenta* (*Anura*: *Hylidae*). *Comp Biochem Physiol, 41A:* 727-730, 1972.

Jungreis, A. M.: Seasonal effects of hyper-osmotic sodium chloride on urea production in the frog, *Rana pipiens. J Exp Zool, 178:* 403-414, 1971.

Kalmus, H.: Über die Natur des Zeitgedächtnisses der Bienen. *Z Vergl Physiol, 20:* 405-419, 1934.

Kalmus, H.: Diurnal rhythms in the axolotl larva and in Drosophila. *Nature, 145:* 72-73, 1940.

Keeton, W. T.: Effects of magnets on pigeon homing. In Galler, S. R., Schmidt-Koening, K., Jacobs, G. J. and Belleville, R. E. (Eds.): *Animal Orientation and Navigation.* Washington, D.C., Scientific and Technical Information Office, National Aeronautics and Space Administration, 1972, pp. 579-594.

Kerfoot, W. B.: The lunar periodicity of *Sphecodogastra texana*, a nocturnal bee (Hymenoptera: Halictidae). *Anim Behav, 15:* 479-486, 1967.

Koltermann, R.: Lern- und Vergessensprozesse bei der Honigbiene -aufgezeight anhand von Duftdressuren. *Z Vergl Physiol, 63:* 310-334, 1969.

Koltermann, R.: 24-Std-Periodik in der Langzeiterinnerung an Duft- und Farbsignale bei der Honigbiene. *Z Vergl Physiol, 75:* 49-68, 1971.

Landreth, H. F. and Ferguson, D. E.: Newts: Sun-compass orientation. *Science, 158:* 1459-1461, 1967.

Lear, J.: The orbiting potato. *Sat Rev, Sept. 4:* 47-51, 1965.

Lickey, M. E.: Seasonal modulation and non-24-hour entrainment of a circadian rhythm in a single neuron. *J Comp Physiol Psychol, 68:* 9-17, 1969.

Lindauer, M.: *Communication Among Social Bees.* Cambridge, Harvard University Press, 1971, 161 pp.

Lindauer, M. and Martin, H.: Die Schwereorientierung der Bienen unter dem Einfluss des Erdmagnetfeldes. *Z Vergl Physiol, 60:* 219-243, 1968.

Loher, W. and Chandrashekaran, M. K.: Circadian rhythmicity in the oviposition of the grasshopper *Chorthippus curtipennis. J Insect Physiol, 16:* 1677-1688, 1970.

McNaughton, S. J.: Enzymic thermal adaptations: The evolution of homeostasis in plants. *Am Nat, 106:* 165-172, 1972.

Madison, D. M. and Shoop, C. R.: Homing behavior, orientation and home range of salamanders tagged with tantalum-182. *Science, 168:* 1484-1487, 1970.

Medugorac, I. and Lindauer, M.: Das Zeitgedächtnis der Bienen unter dem Einfluss von Narkose und von sozialen Zeitgebern. *Z Vergl Physiol, 55*: 450-474, 1967.

Megušar, F.: Experimente über den Farbwechsel der Crustacean. *Arch EntwMech Org, 33*: 462-665, 1912.

Meier, A. H.: Diurnal variations of metabolic responses to prolactin in lower vertebrates. *Gen Comp Endocrinol (Suppl. 2)*: 55-62, 1969.

Meier, A. H., Garcia, L. E. and Joseph, M. M.: Corticosterone phases a circadian water-drive response to prolactin in the spotted newt, *Notophthalmus viridescens. Biol Bull, 141*: 331-336, 1971.

Meier, A. H., John, T. M. and Joseph, M. M.: Corticosterone and the circadian pigeon cropsac response to prolactin. *Comp Biochem Physiol, 40A*: 459-465, 1971.

Menaker, M. (Ed.): *Biochronometry,* Proceedings of a Symposium, 1969, Friday Harbor. Washington, D.C., National Academy of Sciences, 1971, 662 pp.

Menaker, M.: Nonvisual light reception. *Sci Am, 226(3)*: 22-29, 1972.

Menaker, M. and Eskin, A.: Entrainment of circadian rhythms by sound in *Passer domesticus. Science, 154*: 1579-1581, 1966.

Minkiewicz, R.: Rôle des facteurs optiques dans les changements de livrée, chez les Grenouilles adultes. (Etude neurobiologique). *Acta Biol Exp (Vars.) 8*: 102-177, 1933.

Mizell, S.: Seasonal changes in energy reserves in the common frog, *Rana pipiens. J Cell Comp Physiol, 66*: 251-258, 1965.

Morgan, W. W. and Mizell, S.: Diurnal fluctuation in DNA content and DNA synthesis in the dorsal epidermis of *Rana pipiens. Comp Biochem Physiol, 38A*: 591-602, 1971a.

Morgan, W. W. and Mizell, S.: Daily fluctuations of DNA synthesis in the corneas of *Rana pipiens. Comp Biochem Physiol, 40A*: 487-493, 1971b.

Naylor, E.: Tidal and diurnal rhythms of locomotory activity in *Carcinus maenas* (L.). *J Exp Biol, 35*: 602-610, 1958.

Naylor, E., Atkinson, R. J. A. and Williams, B. G.: External factors influencing the tidal rhythm of shore crabs. *J Interdiscipl Cycle Res, 2*: 173-180, 1971.

Naylor, E. and Smith, G.: The role of the eyestalk in the tidal activity rhythm of the shore crab *Carcinus maenas* (L.). In Salánki, J. (Ed.): *Invertebrate Neurobiology: Mechanisms of Rhythm Regulation.* Budapest, Hungarian Academy of Sciences, 1973, in press.

Neher, E.: Clamp currents in the subthreshold voltage range related to pacemaker activity in *Helix* neurons. In Salánki, J. (Ed.): *Invertebrate Neurobiology: Mechanisms of Rhythm Regulation.* Budapest, Hungarian Academy of Sciences, 1973, in press.

Newell, R. C.: Effect of fluctuations in temperature on the metabolism of intertidal invertebrates. *Am Zool, 9*: 293-307, 1969.

Noble, G.: *The Biology of the Amphibia.* New York, McGraw Hill, 1931, (Reprinted, New York, Dover Publishing Co., 1954), 577 pp.

Page, T. L. and Larimer, J. L.: Entrainment of the circadian locomotor activity rhythm in crayfish. The role of the eyes and caudal photoreceptor. *J Comp Physiol, 78:* 107-120, 1972.

Palmer, J. D.: A persistent, light-preference rhythm in the fiddler crab, *Uca pugnax* and its possible adaptive significance. *Am Nat, 98:* 431-434, 1964.

Palmer, J. D.: Daily and tidal components in the persistent rhythmic activity of the crab, *Sesarma. Nature, 215:* 64-66, 1967.

Palmer, J. D. (Ed.): *The Biological Clock. Two Views.* New York and London, Academic Press, 1970, 94 pp.

Papi, F. and Pardi, L.: On the lunar orientation of sandhoppers (Amphipoda, Talitridae). *Biol Bull, 124:* 97-105, 1963.

Pavlidis, T.: Mathematical models of circadian rhythms: Their usefulness and their limitations. In Menaker, M. (Ed.): *Biochronometry.* Washington, D. C., National Academy of Sciences, 1971, pp. 110-116.

Percy, J. A. and Aldrich, F. A.: Metabolic rate independent of temperature in mollusc tissue. *Nature, 231:* 393-394, 1971.

Pittendrigh, C. S.: On temperature independence in the clock system controlling emergence time in *Drosophila. Proc Nat Acad Sci, 40:* 1018-1029, 1954.

Pittendrigh, C. S.: Circadian rhythms and the circadian organization of living systems. *Cold Spring Harbor Symp Quant Biol, 25:* 159-184, 1960.

Pittendrigh, C. S.: Photoperiodism and biological clocks. In Moore, J. A. (Ed.): *Proc XVI Internat Congress Zool.* Washington, D. C., XVI International Congress of Zoology, 1963, vol. IV, p. 360.

Pittendrigh, C. S. and Bruce, V. G.: An oscillator model for biological clocks. In Rudnick, D. (Ed.): *Rhythmic and Synthetic Processes in Growth.* Princeton, Princeton University Press, 1957, pp. 75-109.

Pittendrigh, C. S. and Bruce, V. G.: Daily rhythms as coupled oscillator systems and their relation to thermoperiodism and photoperiodism. In Withrow, R. B. (Ed.): *Photoperiodism and Related Phenomena in Plants and Animals.* Washington, D. C., American Association for the Advancement of Science, 1959, pp. 475-505.

Pittendrigh, C. S. and Minis, D. H.: The photoperiodic time measurement in *Pectinophora gossypiella* and its relation to the circadian system in that species. In Menaker, M. (Ed.): *Biochronometry.* Washington, D. C., National Academy of Sciences, 1971, pp. 212-250.

Presman, A. S.: *Electromagnetic Fields and Life,* (F. L. Sinclair, trans.). New York and London, Plenum Press, 1970, 336 pp.

Ralph, C. L.: Persistent rhythms of activity and O_2 consumption in the earthworm. *Physiol Zool, 30:* 41-55, 1957a.

Ralph, C. L.: A diurnal activity rhythm in *Plethodon cinereus* and its modification by an influence having a lunar frequency. *Biol Bull, 113*: 188-197, 1957b.

Rao, K. P.: Tidal rhythmicity of rate of water propulsion in *Mytilus,* and its modifiability by transplantation. *Biol Bull, 106*: 353-359, 1954.

Reinke, E. E. and Chadwick, C. S.: The origin of the water drive in *Triturus viridescens.* I. Induction of the water drive in thyroidectomized and gonadectomized land phases by pituitary implantations. *J Exp Zool, 83*: 223-233, 1940.

Renner, M.: Über die Haltung von Bienen in geschlossenen künstlich beleuchteten Räumen. *Naturwissenschaften, 42*: 539-540, 1955a.

Renner, M.: Ein Transozeanversuch zum Zeitsinn der Honigbiene. *Naturwissenschaften, 42*: 540-541, 1955b.

Renner, M.: Neue Versuche über den Zeitsinn der Honigbiene. *Z Vergl Physiol, 40*: 85-118, 1957.

Renner, M.: Über ein weiteres Versetzungsexperiment zur Analyse des Zeitsinnes und der Sonnenorientierung der Honigbiene. *Z Vergl Physiol, 42*: 449-483, 1959.

Renner, M.: Zeitsinn und astronomische Orientierung der Honigbiene. *Naturwiss Rdsch, 14*: 296-305, 1961.

Röseler, I.: Die Rhythmik der Chromatophoren des Polychaeten *Platynereis dumerilii. Z Vergl Physiol, 70*: 144-174, 1970.

Russell, D. R.: Effect of a constant magnetic field on invertebrate neurons. In Barnothy, M. F. (Ed.): *Biological Effects of Magnetic Fields.* New York and London, Plenum Press, 1969, vol. II, pp. 227-232.

Salánki, J. (Ed.): *Invertebrate Neurobiology: Mechanisms of Rhythm Regulation.* Proceedings of a Symposium, 1971, Tihany. Budapest, Hungarian Academy of Sciences, 1973, in press.

Salánki, J. and Vero, M.: Diurnal rhythm of activity in fresh-water mussel *(Anodonta cygnea L.)* under natural conditions. *Annals of Biology, Tihany, 36*: 95-107, 1969.

Sandeen, M. I., Stephens, G. C. and Brown, F. A., Jr.: Persistent daily and tidal rhythms of oxygen consumption in two species of marine snails. *Physiol Zool, 27*: 350-356, 1954.

Sauer, E. G. F. and Sauer, E. M.: Star navigation of nocturnal migrating birds. *Cold Spring Harbor Symp Quant Biol, 25*: 463-473, 1960.

Scharrer, E. and Brown, S.: Neurosecretion. XII. The formation of neurosecretory granules in the earthworm, *Lumbricus terrestris L. Z Zellforsch, 54*: 530-540, 1961.

Schauble, M. K.: Seasonal variation of newt forelimb regeneration under controlled environmental conditions. *J Exp Zool, 181*: 281-286, 1972.

Scheving, L. E., Halberg, F. and Pauly, J. E. (Eds.): *Chronobiology.* Proceedings of the International Society for the Study of Biological Rhythms, 1971, Little Rock, Ark., Tokyo, Igaku Shoin, Ltd., 1973, in press.

Simpson, J. A.: Cosmic-radiation intensity-time variations and their origin. III. The origin of 27-day variation. *Physical Rev, 94*: 426-440, 1954.

Sollberger, A.: *Biological Rhythm Research.* Amsterdam-London-New York, Elsevier Publ. Co., 1965, 461 pp.

Southern, W. E.: Influence of disturbances in the earth's magnetic field on ring-billed gull orientation. *Am Zool, 10*: 292-293, 1970.

Southern, W. E.: Magnets disrupt the orientation of juvenile ring-billed gulls. *BioScience, 22*: 476-479, 1972.

Stephens, G. C.: Influence of temperature fluctuations on the diurnal melanophore rhythm of the fiddler crab, *Uca. Physiol Zool, 30*: 55-69, 1957.

Stephens, G. C.: Circadian melanophore rhythms of the fiddler crab: Interaction between animals. *Ann N Y Acad Sci, 98*: 926-939, 1962.

Stephens, G. C., Sandeen, M. I. and Webb, H. M.: A persistent tidal rhythm of activity in the mud snail, *Nassa obsoleta. Anat Rec, 117*: 635, 1953.

Stephens, G. J., Halberg, F. and Stephens, G. C.: The blinded fiddler crab: An invertebrate model of circadian desynchronization. *Ann N Y Acad Sci, 117*: 386-406, 1964.

Steven, D. M.: The dermal light sense. *Biol Rev, 38*: 204-240, 1963.

Strumwasser, F.: The demonstration and manipulation of a circadian rhythm in a single neuron. In Aschoff, J. (Ed.): *Circadian Clocks.* Amsterdam, North-Holland Publ. Co., 1965, pp. 442-462.

Strumwasser, F.: Types of information stored in single neurons. In Wiersma, C. A. G. (Ed.): *Invertebrate Nervous Systems.* Chicago, University of Chicago Press, 1967, pp. 291-319.

Sweeney, B. M.: *Rhythmic Phenomena in Plants.* London and New York, Academic Press, 1969, 147 pp.

Sweeney, B. M. and Hastings, J. W.: Effects of temperature upon diurnal rhythms. *Cold Spring Harbor Symp Quant Biol, 25*: 87-104, 1960.

Szymanski, J. S.: Eine Methode zur Untersuchung der Ruhe- und Aktivitäts-perioden bei Tieren. *Pfluger's Arch Ges Physiol, 158*: 343-385, 1914.

Szymanski, J. S.: Die Verteilung von Ruhe- und Aktivitätsperioden bei einigen Tierarten. *Pfluger's Arch Ges Physiol, 172*: 430-448, 1918.

Taylor, D. H.: Biological clock control and extraoptic photoperception in the tiger salamander, *Ambystoma tigrinum. Am Zool, 10*: 474, 1970.

Taylor, D. H. and Ferguson, D. E.: Extraoptic celestial orientation in the southern cricket frog, *Acris gryllus. Science, 168*: 390-392, 1970.

Twitty, V. C.: *Of Scientists and Salamanders.* San Francisco and London, W. H. Freeman and Co., 1966, 178 pp.

von Buttel-Reepen, H.: *Leben und Wesen der Bienen.* Braunschweig, Vieweg und Sohn, 1915, 300 pp.

von Frisch, K.: *Bees: Their Vision, Chemical Senses and Language.* Ithaca, Cornell University Press, 1971, 157 pp.

von Frisch, K.: *The Dance Language and Orientation of Bees,* (Chadwick,

L. E., trans.). Cambridge, Belknap Press of Harvard University Press, 1967, 566 pp.

Wahl, O.: Neue Untersuchungen über das Zeitgedächtnis der Bienen. Z *Vergl Physiol, 16*: 529-589, 1932.

Webb, H. M.: Diurnal variations of response to light in the fiddler crab, *Uca. Physiol Zool, 23*: 316-337, 1950.

Webb, H. M.: Pigmentary rhythms as indicators of neurosecretion. *Am Zool, 6*: 181-186, 1966.

Webb, H. M.: Effects of artificial 24-hour cycles on the tidal rhythm of activity in the fiddler crab, *Uca pugnax. J Interdiscipl Cycle Res, 2*: 191-198, 1971.

Webb, H. M., Bennett, M. F. and Brown, F. A., Jr.: A persistent diurnal rhythm of chromatophoric response in eyestalkless *Uca pugilator. Biol Bull, 106*: 371-377, 1954.

Webb, H. M. and Brown, F. A., Jr.: The repetition of pattern in the respiration of *Uca pugnax. Biol Bull, 115*: 303-318, 1958.

Webb, H. M. and Brown, F. A., Jr.: Seasonal variations in O_2-consumption of *Uca pugnax. Biol Bull, 121*: 561-571, 1961.

Webb, H. M. and Brown, F. A., Jr.: Interactions of diurnal and tidal rhythms in the fiddler crab, *Uca pugnax. Biol Bull, 129*: 582-591, 1965.

Webb, H. M., Brown, F. A., Jr. and Brett, W. J.: Fluctuations in rate of locomotion in *Ilyanassa. Biol Bull, 117*: 431, 1959.

Wever, R.: Über die Beeinflussung der circadianen Periodik des Menschen durch schwache elektromagnetische Felder. Z *Vergl Physiol, 56*: 111-128, 1967.

Wever, R.: Influence of electric fields on some parameters of circadian rhythms in man. In Menaker, M. (Ed.): *Biochronometry*. Washington, D. C., National Academy of Sciences, 1971, pp. 117-133.

Wieser, W., Fritz, H. and Reichel, K.: Jahreszeitliche Steuerung der Atmung von *Arianta arbustorum* (Gastropoda). Z *Vergl Physiol, 70*: 62-79, 1970.

Wiltschko, W.: Über den Einfluss statischer Magnetfelder auf die Zugorientierung der Rotkehlchen. Z *Tierpsychol, 25*: 537-558, 1968.

Winfree, A. T.: Corkscrews and singularities in fruitflies: Resetting behavior of the circadian eclosion rhythm. In Menaker, M. (Ed.): *Biochronometry*. Washington, D. C., National Academy of Sciences, 1971, pp. 81-109.

AUTHOR INDEX

A

Abramowitz, A. A., 22, 23, 24, 36, 183
Adler, K., 159, 160, 162, 163, 183
Ajrapetyan, S. N., 128, 183
Aldrich, F. A., 179, 192
Arbit, J., 101, 183
Aréchiga, H., 71, 183
Arudpragasam, K. D., 70, 183
Ashoff, J., 7, 13, 17, 29, 93, 95, 164, 183
Atkinson, R. J. A., 66, 191
Axelrod, J., 160, 183

B

Bachman, C. H., 121, 188
Baldwin, F. M., 100, 101, 183
Barnett, A., 152, 183
Barnothy, M. F., 120, 183, 184
Barnwell, F. H., 43, 53, 63, 130, 131, 134, 184
Barrera, B., 71, 183
Beck, S., 184
Becker, R. O., 121, 188
Beier, W., 80, 88, 90, 184
Beling, I., 76, 77, 80, 83, 91, 184
Bennett, M. F., 34, 36, 38, 40, 44, 54, 58, 60, 64, 89, 102, 103, 104, 106, 108, 110, 111, 112, 113, 114, 115, 116, 117, 118, 119, 121, 128, 131, 132, 135, 136, 141, 162, 164, 184, 185, 186, 187, 195
Bentley, P. J., 153, 185
Bliss, D., 64, 69, 185
Bohn, G., 124, 127, 185
Brady, J., 115, 185
Bräuninger, H. D., 96, 185
Brett, W. J., 137, 138, 187, 195
Brinkmann, K., 180, 185
Brown, F. A., Jr., 7, 10, 12, 19, 20, 24, 25, 29, 30, 33, 34, 36, 37, 38, 40, 44, 45, 53, 54, 58, 60, 62, 64, 71, 108, 109, 121, 128, 129, 131, 132, 135, 136, 137, 138, 139, 141, 143, 144, 164, 173, 175, 176, 180, 184, 185, 186, 187, 193, 195
Brown, R. A., 58, 185
Brown, S., 113, 193
Bruce, V. G., 173, 192
Bünning, E., 6, 7, 16, 173, 187
Burkhardt, D., 179, 187

C

Chadwick, C. S., 169, 193
Chandrashekaran, M. K., 160, 190
Chapman, S., 50
Cloudsley-Thompson, J. L., 164, 187
Cole, L. C., 52, 187
Conroy, R. T. W. L., 187
Crane, J., 21, 43, 187

D

Darwin, Charles, 4, 15, 16, 98, 99, 100, 101, 122, 159, 187
Daumer, 86
De Candolle, 9
DeCoursey, P. J., 13, 15, 187
Dodd, J. R., 134, 187
Duhamel, 4, 5
Dumortier, B., 160, 188
Dunlap, D. G., 167, 188

E

Ehret, C. F., 150, 152, 183, 188
Eiseley, L., 3, 188
Emlen, S. T., 120, 188

197

SUBJECT INDEX